# PC & Electronics

## Connecting Your PC to the Outside World

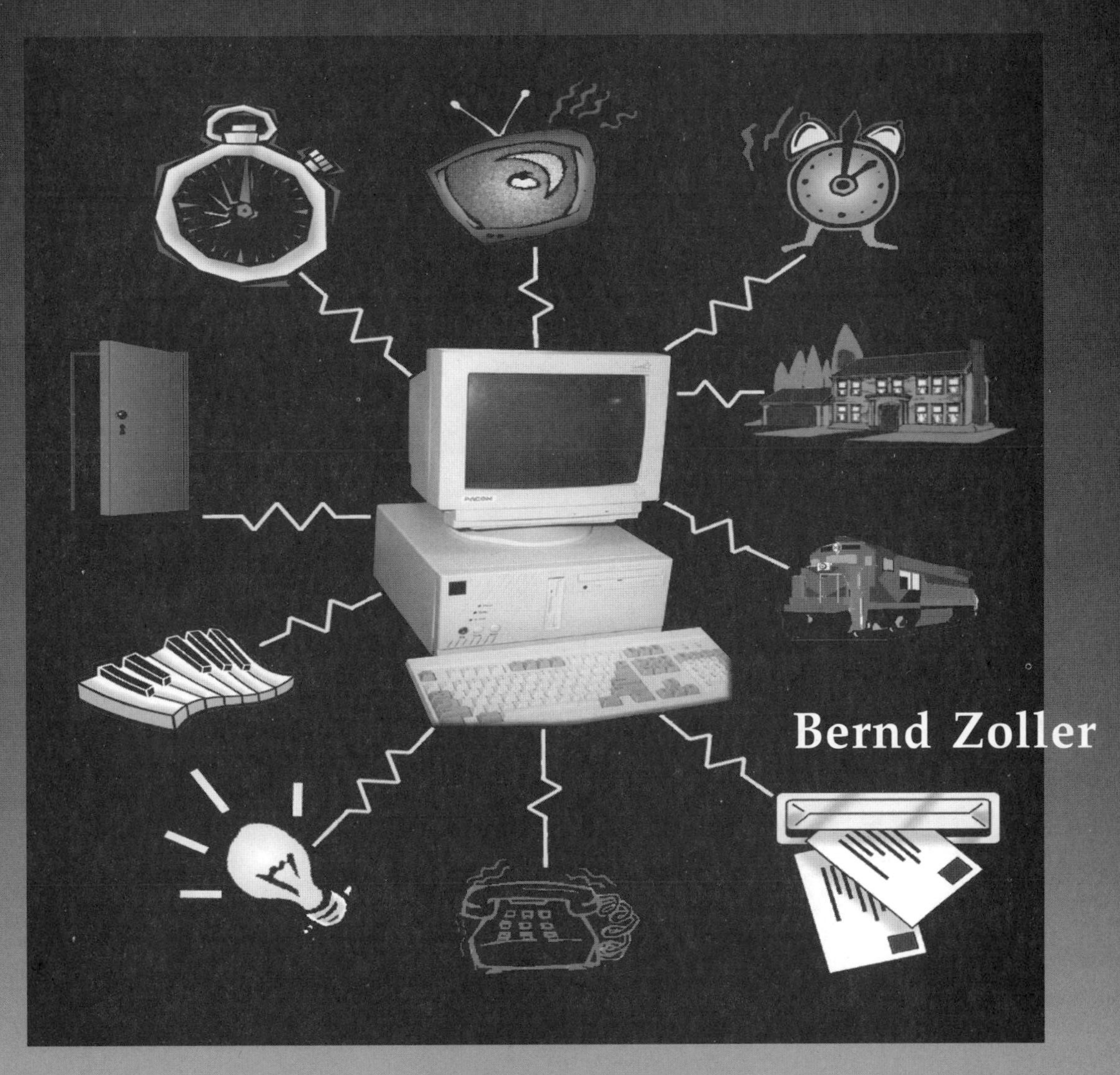

Bernd Zoller

DATA BECKER

Abacus

Printed in the U.S.A.

ISBN 1-55755-334-3

10 9 8 7 6 5 4 3 2 1

# Contents

# Chapter 1: Introduction

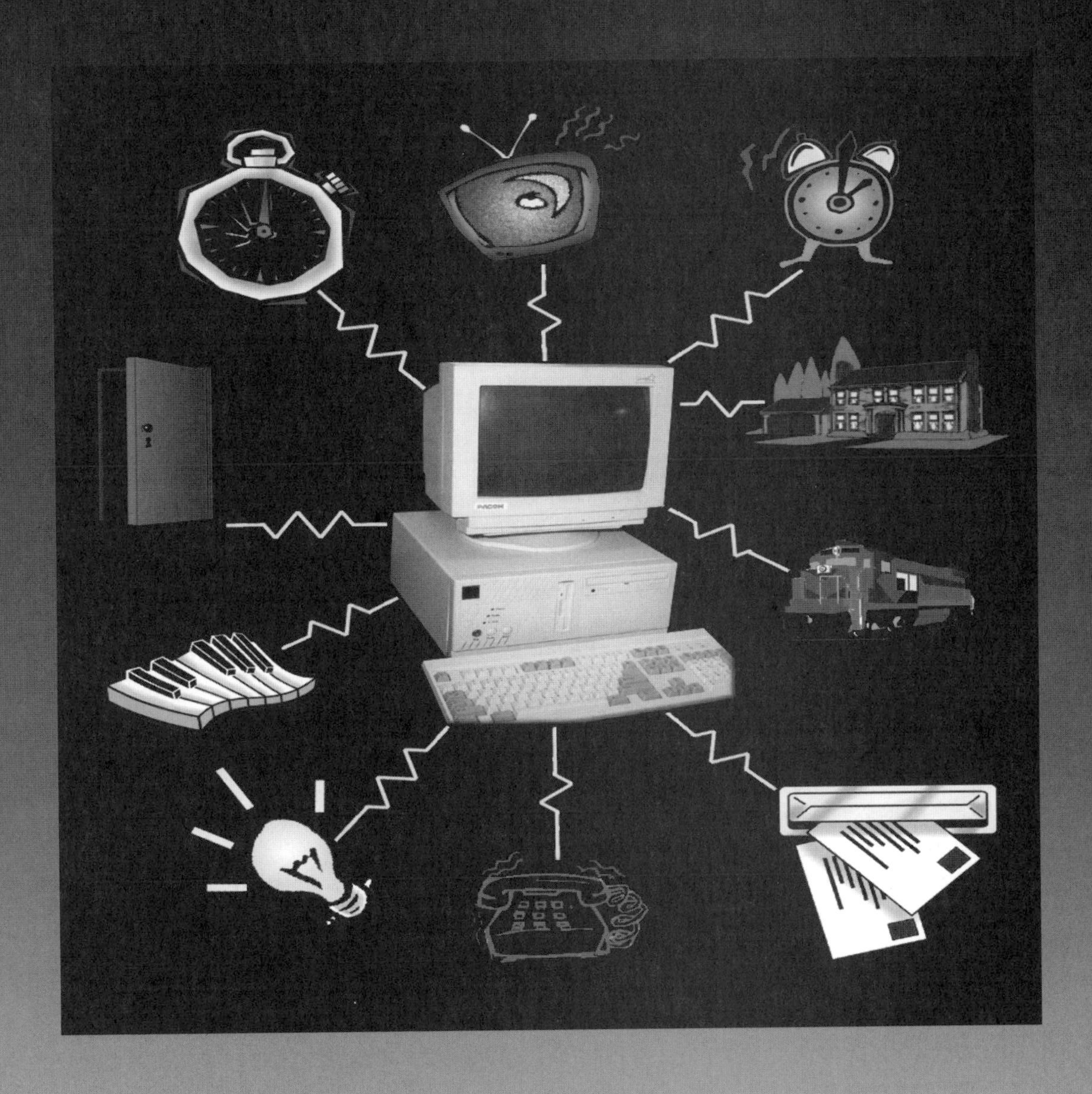

# Chapter 1: Introduction

Because computers can accomplish many tasks, their range of application is correspondingly diverse. Quite often in the business world, employees use computers for writing or communication. Many companies use the computer as a developmental tool or as a control device for production routines.

On the other hand, computers often have different purposes at home. The only tasks that home PCs have in common with business PCs are the writing of letters or managing personal data. Computers are frequently used at home for entertainment purposes. However, this doesn't have to remain their only use at home. Often, you can also use the computer in housekeeping or with a hobby. This book explores this idea and gives you several options for putting this into practice.

## The Basic Idea Of This Book

Basically, we do not approach the computer as a data management device in this book. Instead, we'll show you how to use your computer as a control device. For example, we won't tell you how to use your computer to manage your stamp collection. We will, however, show you how to use your computer as a remote control device for a model train.

The big difference between the two examples we just mentioned is that the computer in the first example remains limited to managing the software. You input data for the stamp collection and the software performs the desired actions within the computer.

The second example is different. Although the second example also uses software, the effects do not stay inside the computer. Instead, the results of using the software will affect the environment of the computer.

However, as a rule you can only transfer these effects to the outside world by connecting the environment and your computer. In most cases, personal computers have one or more such connection options, used almost exclusively for specific purposes. You take advantage of these yourself daily, although you often don't give them a second thought. These connection options are the ports of your computer.

## Parallel and serial ports

The parallel port is primarily used to connect your computer to a printer. Most PCs have two serial ports; these are used to connect the computer to a mouse or modem.

However, it doesn't have to stay this way. The computer can output or receive all types of data through the ports. There are two requirements for meaningful data exchange between the computer and the environment. One is that the world outside the computer must be able to properly process and interpret the data the computer outputs. The other requirement is that data the computer receives from the outside world must be prepared so that the computer can process it.

Both of these requirements can only be fulfilled by connecting additional hardware to the ports for performing the necessary conversions. As our book shows, this doesn't necessarily have to be expensive and complicated.

## The purpose of this book

We wrote this book to give you the opportunity to use your computer to accomplish specific tasks, either at home or in connection with a hobby. To make this happen, you'll need to connect additional hardware to one of your computer ports.

However, don't be intimidated by this complicated sounding description. The aim of this book is to furnish you with information so that you'll be able to create the necessary circuits for the intended purpose yourself, without expensive investments. The chapters of this book will teach you how to create the circuits and what you can accomplish with them.

## For Whom Is This Book Written?

You've just learned that this book will guide you in creating your own circuits to establish a connection between your computer and your technical surroundings. Now, you may be thinking that you don't have the necessary technical background to build the electronic circuits. However, don't be too hasty in your judgment. First read the following paragraphs, which describe the technical requirements.

### The technical requirements

Before we tell you what electronic knowledge you must have to build or create the circuits, we'll explain which technical prerequisites are necessary.

Basically, the circuits we'll talk about in this book should be operable using a standard IBM-compatible PC. It doesn't matter whether you use an "old" 386 or a fast Pentium. It simply needs to have one parallel and one serial port. Also, Windows 3.1 or a higher version of Windows must be installed on your PC.

To create the circuits, you have to put together some electronic components. However, it's not necessary for you to be familiar with the operation of these electronic components. The chapters will give you all the information you need to know to put the circuits together. In most cases, you'll build the circuits on an all-purpose substrate. You can easily find this and all the other electronic components at an electronics shop.

In other words, you don't have to be able to "read" and understand circuit diagrams. Furthermore, the explanations and instructions are written in easy to understand language; you won't be confronted with technical terminology.

### The required skills

To create the circuits, it is necessary to perform specific electronic tasks. However, these are limited to operations that you can easily reproduce or can learn with a little practice. Typically, you'll be using a soldering iron, side-cutting pliers and similar tools. Acquiring the necessary skill for using the tools properly will be easy.

We explain all the necessary tasks in easy to understand language. However, it's going to take some time before you'll be able to perform the necessary tasks easily. Remember the old motto: Practice makes perfect. If you have any kind of background in this area, you shouldn't have any trouble at all understanding what we describe in this book.

So, if you are interested in creating simple circuits that will connect your computer to its technical surroundings, read this book. It gives you everything you need to convert and use the circuits. The book goes into great detail, and is written in easy to understand layperson's language.

## How The Book Is Structured

The book is divided into three parts. Each part deals with a specific area. The following summarizes the different parts:

### Part I

The first part of the book talks about circuit fundamentals and the required tasks. You get information about electronic components, what they are used for and how they are installed in the circuits. You also get basic information about the required steps for building the circuits. Hints and tips on using the tools complete the information.

To give you a better understanding of the circuit functions, Part I also has a chapter that deals with the functions of the serial and parallel ports. In the explanations, as in all the chapters of this book, we purposely avoid the deep technical jargon, so that even laypersons can make use of the information.

Finally, we discuss the software you'll need for the individual circuits. Here again, we limit our explanations to the essentials.

### Part II

Part II of the book contains the circuits. We discuss each circuit in its own chapter. The circuits are simple in function and structure, so that even beginners can build and use them.

We describe the circuits in Part II so comprehensively that you'll be able to carry out the necessary steps for putting them together. Exact, step-by-step instructions guarantee you'll create the circuits in such a way that you'll be able to use them without any difficulties.

### Part III

Part III goes a bit beyond the beginner level. While laypersons could build these circuits using the instructions, the necessary steps for building them are a bit more complex.

Don't try to build these circuits until you have had enough experience with the circuits in Part II.

## The Companion CD-ROM

Along with the technical and mechanical explanations on the circuits we've touched upon, you'll also need to use software. Since you'll need to use a separate software program for each of the circuits, we have included a companion CD-ROM which contains all the software for each of the circuits. This doesn't just include the executable files, but also the entire source code, so that you can make changes to the source code if necessary.

The structure of the CD-ROM follows the organization of the circuits in the book. Thus, each circuit, which is described in its own chapter in the book, also has its own folder on the CD-ROM. For example, the software for the circuit described in Chapter 20 will be on the CD-ROM in the folder named *CHAP_20*.

With a few exceptions, the folders for the chapters of the book contain subfolders. In most cases, there is a subfolder named *Windows*. This subfolder contains the software for the circuits both in installable and modifiable form.

You will find more information about the contents of the subfolders in the individual chapters. Therefore, pay attention to the hints and comments in the chapters of this book. For more information about the software, see Chapter 6, "The Software for the Circuits."

# Part I: The Basics

Part I of the book gives you basic information about the electronic components used in this book and the tasks to be performed, such as soldering or preparation of materials.

You also get information about the required tools. Hints and tips on using these tools complete Part I.

We also discuss how the serial and parallel ports operate, as well as their differences. This makes it easier for you to understand how the exchange of information or the flow of control occurs.

Since hardly a circuit is able to operate without software, we discuss this in another chapter of the book. In this chapter you'll learn a bit about the programming language we used and your options for modifying the default software.

All in all, Part I of this book gives you the preparation you need to manage Parts II and III of the book with no trouble.

# Chapter 2:

# The Required Tools

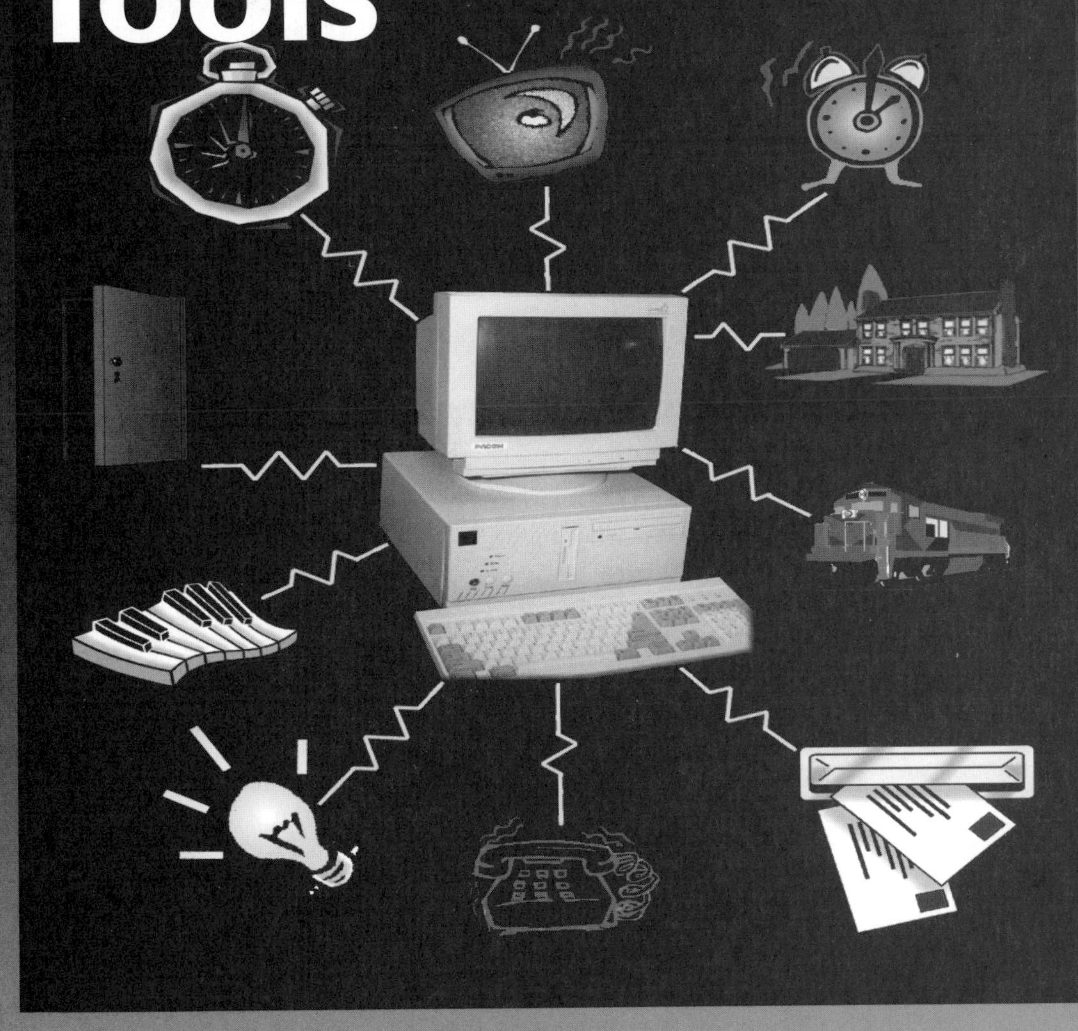

# Chapter 2: Required Tools

As you read in Chapter 1, this book offers ideas and instructions for circuits that you can use with your computer. Although this book includes a companion CD-ROM describing the circuits, it does not include the circuits. You'll have to create the circuits yourself.

Each chapter gives you exact descriptions for building the circuits. However, before you begin, you'll need to be familiar with the tools necessary for building the circuits. Fortunately, you'll only need a few tools to perform the work and no expensive or special tools are needed. You probably already have many of these tools so you won't even need to buy them. In this chapter we'll show which tools you are going to need to build all the circuits. We'll briefly discuss each tool's purpose and show you how to use it.

## Selecting The Right Screwdrivers

You're probably already familiar with using a screwdriver. But there are many types of screwdrivers that you'll probably use and we'll talk about those in this section.

### Flathead screwdriver

You'll need a small to medium-sized flathead screwdriver for working with the circuits. This type of screwdriver has a flat blade that fits into the slot of the appropriate screws (see illustration on the right).

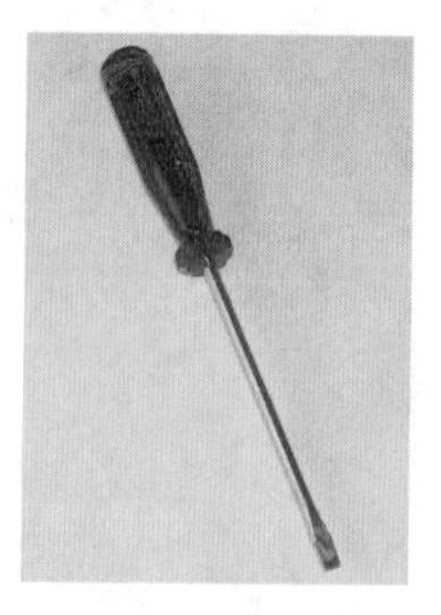

Use this type of screwdriver whenever you have to tighten fastening screws. For example, you may need to do this with plug cases around which you want to attach a plug or if you need to install a plug on a port. As a rule, these examples will use small screws that you can tighten with a flathead screwdriver.

These screwdrivers have several purposes when you are building circuits. For example, a screwdriver is useful when you're soldering. You'll be repeatedly pressing components in a specific position near the soldering joint. Because of the heat given off by a soldering iron, a screwdriver is a better choice for this than your fingers.

**Important Tip**

When you work with screwdrivers, remember to use a screwdriver that matches the size of the screw. Otherwise, you run the risk of "stripping" the screw and damaging it to where you can no longer use it.

We recommend using a flathead screwdriver. However, make certain that it's not too big; instead use one that is handy for this type of work.

### Phillips-head screwdriver

Besides a flathead screwdriver, you'll also find a medium-sized Phillips-head screwdriver to be very useful. This screwdriver doesn't have a flat blade at the end. Instead, it comes to a point and fits screws with heads shaped like a plus sign.

## Cutting Pliers

Cutting pliers are the tool you'll be using most frequently in your work with the circuits.

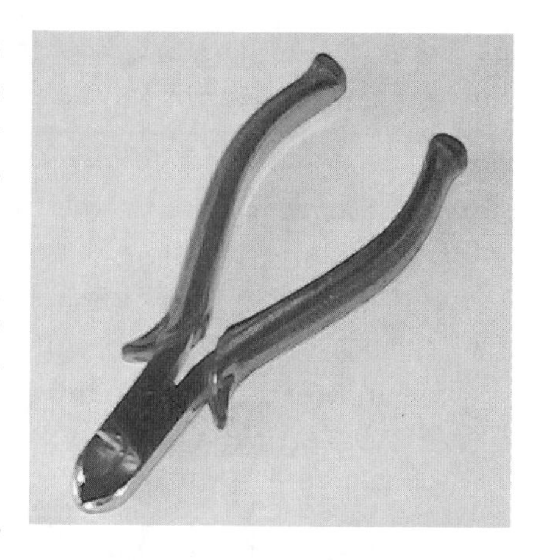

You'll use the cutting pliers often (usually to cut wires). They look like normal pliers but instead of a gripping surface at the end, they have two short cutters. So, as a result of the great leverage, they're able to cut wires.

Because you'll need to cut both thick and thin wires, stores usually sell sturdy, large cutting pliers and small, handy cutting pliers. The big ones are electrical cutting pliers, which will already be standard equipment in some households. The smaller models are cutting pliers specially designed for use in the field of electronics. For working with the circuits you will require a small pair of cutting pliers. While you could also use the smaller models of the electrical cutting pliers, you won't be able to manage so easily with them. Especially when it's a matter of cutting the thin wires precisely, the big cutting pliers aren't suitable.

**Important Tip**

If you decide to purchase a new pair of cutting pliers, be sure to read the information in the following sections. Often electronics stores have complete sets of electronics tools available at reasonable prices. You don't have to buy expensive special models to perform the tasks described in this book.

## Different Types Of Pliers

Along with a pair of cutting pliers, you will need one or two other types of pliers. Here again, we distinguish between the large electrical pliers and the fine electronic pliers.

## Electronic pliers

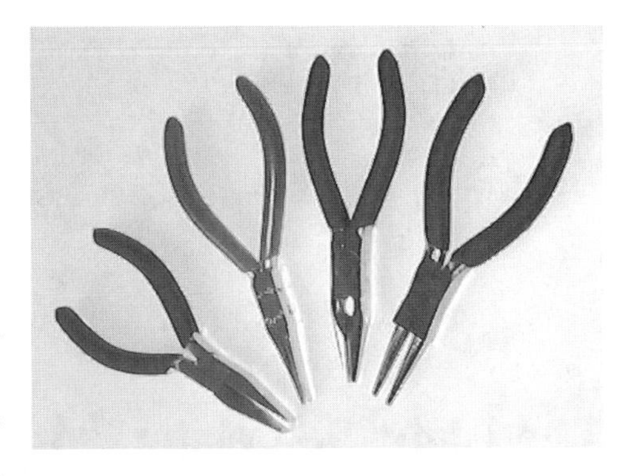

The electronic pliers are absolutely essential. For one thing, you will need pliers to position the electronic components on the board. For another, in creating the wire connections between the components, you will have to bend and adjust fine wires as precisely as possible, in most cases, by 90 degrees. The big electrical pliers are not suitable for this work. Often their ends aren't flat or pointed enough to work properly between the wires. Also, it is often impossible to grab onto the fine components and wires with these clumsy pliers.

## Flat-nosed pliers

You will need flat-nosed pliers to align and position the wires and components. Flat-nosed pliers have two very flat jaws, or gripping wedges, at their ends, so you can use them very effectively in narrow gaps, for example, between wires.

You can use these pliers to bend and adjust fine wires. While you cannot exert great force with these pliers, especially not at the outer end (for example, to crimp something together), you won't need to use great force with the flat-nosed pliers anyway.

## Long-nose pliers

Besides the flat-nosed pliers, you'll probably often use long-nose pliers. Although similar in shape to flat-nosed pliers, the jaws (or gripping wedges) on long-nose pliers are narrower. Also, the jaws of long-nose pliers are sturdier in design. This is often achieved by forming the outer jaw sides in the shape of a semi-circle. You can exert a bit more pressure with these pliers on the components. Moreover, it is often easier to bend wires with these pliers, especially when you need to have several bends near each other.

**Important Tip**

You can usually buy flat-nosed pliers, long-nose pliers and cutting pliers as a set for a reasonable price. The tools in this set will satisfy the requirements of your hobby. Professional electronic tools, some of which cost more per tool than the entire hobby set we've been discussing, are not necessary for the tasks in this book.

## The Right Soldering Iron

In addition to the tools we've already mentioned, you'll also need a soldering iron. A soldering iron is a necessity; you cannot build the circuits without one.

You may already own a soldering iron; if so it will probably be good enough for the work you need to do here. However, you could have a problem if your soldering iron has been designed for more "industrial" work, such as hard soldering of sheet metal instead of precision electronics soldering work. In such a case, read through the following paragraphs to decide if you need to get a more delicate tool.

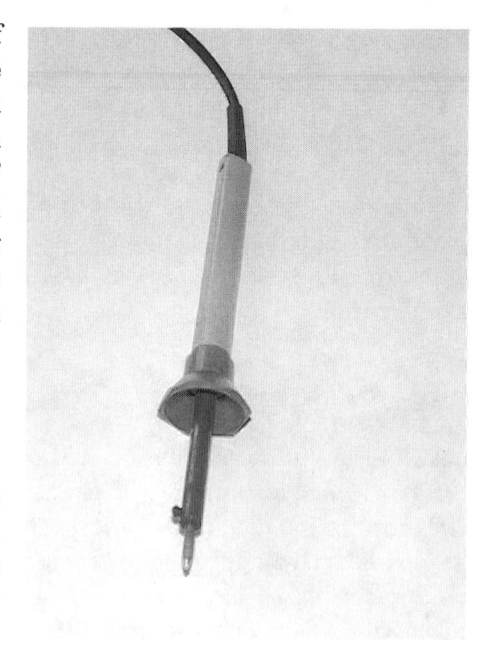

### How a soldering iron works

The soldering iron uses solder to create electrical connections between components. The components become so hardened by the solder that they retain their position, even when slight mechanical pressure is exerted on them. In this process, the soldering iron heats the solder until it becomes a liquid and can create the connection.

To accomplish this task, electrical heating elements are activated within the soldering iron, which generate the necessary heat. The soldering tasks we talk about here don't require extremely high temperatures. Therefore, relatively simple soldering irons with a power level of approximately 30 watts will work fine. This type of soldering iron costs anywhere from 5 to 20 dollars.

**Important Tip**

Using a soldering iron that's too powerful can create problems. The solder can easily overheat and burn so it cannot be processed properly. In addition, sensitive components can be damaged by the excessive heat.

## The soldering tip

Along with the heating element, the soldering tip is another important part of the soldering iron. Heat is transferred to the solder and the component through the soldering tip. Soldering iron tips are usually interchangeable, though there are different types and shapes of soldering tips. The simplest kind is made of copper and has a shape similar to the tip of a flathead screwdriver. The heat is transferred to the soldering joint through the flat end of the soldering tip.

Round tips are better suited for precision soldering work than the flat tips. It's easier to make contact with the desired soldering joint using round tips. This ability will make your soldering work easier. It is always frustrating when a soldering connection next to the actual soldering joint dissolves because you unintentionally heated it.

Another difference lies in the quality of the soldering tip. Tips are made either entirely of copper, nickle-plated, or are covered with another durable substance. They have the advantage of being easier to clean. Copper tips sometimes require a wire brush or a file to remove the deposits from heat emission. The nickel-plated tips, on the other hand, can be heated and easily cleaned with a cloth. Also, the copper tips have the disadvantage of wearing out after a period of time so that it becomes occasionally necessary to remachine or dress the shape of the tip. The improved tips don't have this disadvantage. Since the coated soldering tips are relatively inexpensive, we recommend buying one if your soldering iron is not already equipped with such a tip.

## Soldering holder

When working with the soldering iron, it's important that you're able to place it on the work table in a fixed position. If you simply set the soldering iron on the table, it could easily fall to the floor. This could result in slight burns or a singed area on your carpeting. As a safety precaution, you need to get a soldering iron holder. You can easily set the soldering iron down in this holder. The fixed location of the holder on the table makes it much harder for the soldering iron to slip.

## Using a sponge to clean the tips

The development of heat will cause residue to form repeatedly on the tip of the soldering iron. Before soldering, you need to remove this residue, so that only "fresh" solder goes on the soldering joint.

The easiest way to keep the soldering iron tip clean is to use a small, damp sponge. It's best if you lay the sponge flat on the table, or even better, combine it with the soldering iron holder. That holds the sponge in one place so that it cannot slip.

**Important Tip**

Many electronics suppliers also have starter kits available. These kits usually include everything you will need for soldering and are good enough for the tasks we talk about here. These kits may also include cutting pliers, flat-nosed pliers and long-nose pliers. If so, you can buy most of the tools you will need in one set at a reasonable price.

## Additional Tools To Have Handy

You can build all the circuits we'll talk about in this book using just the tools we have described so far. However, we'll talk about some other tools in this section that will make your work even easier. Don't panic if you don't have these tools. You won't need to buy them right now; wait until you get some more experience. If you discover that these tools will make the work easier, you can always buy them later.

### Air-suction nozzle

If you don't have much experience in soldering, your soldering joints may be a little less than perfect. In such a case, you will often have to free the soldered joint from the solder. For this you can use an air-suction nozzle.

The air-suction nozzle helps you remove solder. At the push of a button, this nozzle creates a vacuum that "swallows" the excess fluid solder. Then you can begin working on the soldered joint again.

**Important Tip**

You can also use a solder wick for removing solder. This is a multiconductor copper cable that you hold on the liquid solder. The solder wick picks up the solder. This method of removing solder is not as efficient as the air-suction nozzle.

### The card cage

After you have done some soldering, there will be times when you probably wish you had a third or maybe even a fourth hand. In one hand you hold the soldering iron, in the other hand, the solder. It would be good if the soldering joint, e.g., the component to be soldered and the substrate, could remain on the table in one location that is easy to access. However, often this won't be the case. Frequently, a card cage can be very helpful. This is a device that can be placed or screwed firmly to a flat surface, for example, a table. Using small grip arms, you can hold the board in place so that the soldering joint is easily accessible and won't be moved out of position by the slight pressure of the soldering iron.

**Important Tip**

A small vise can also act as a third hand. It would be of great advantage if you were able to freely position the vise using a ball joint. If you have a vise, it will certainly make your work easier.

## The forceps

If you have ever looked at an electronic circuit, for example, in a defective radio, you will already be familiar with the dimensions of the electronic components. If not, then you will soon become acquainted with them. In any case, some of the components are quite small, making them difficult to grab with your fingers. In addition, these components must often be installed in hard-to-reach places. In such cases, forceps can be very helpful. It doesn't matter whether the forceps are straight or curved. If you own forceps, set them aside for your soldering work. If you don't own a pair, first see if you can manage with the pliers.

# Chapter 3:

# Necessary Tasks

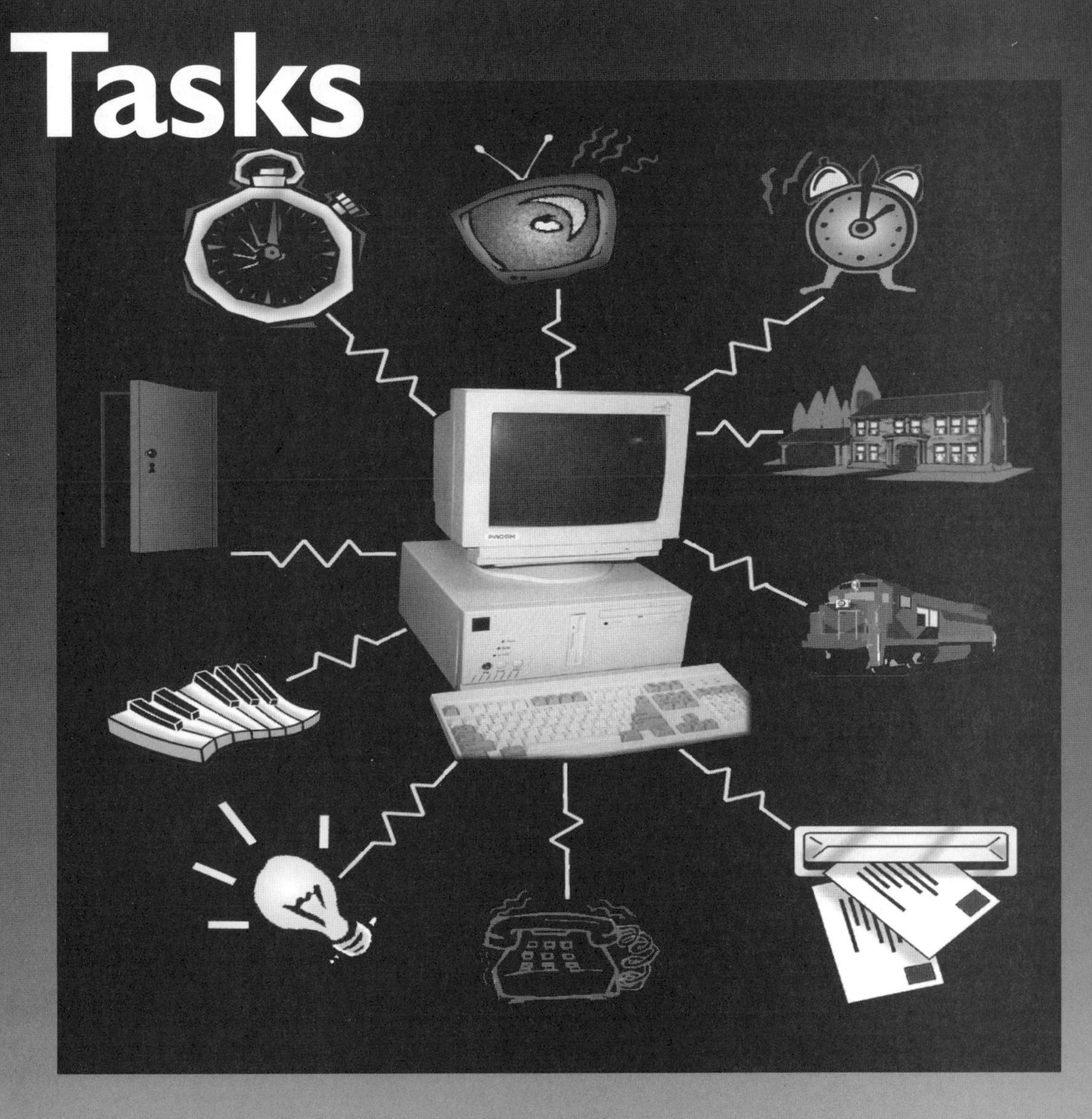

# Chapter 3: Necessary Tasks

In this book we introduce you to several circuits that you can build. You don't have to understand the functional routines within the circuit to do this. In other words, you don't have to be an electronics engineer to use the circuits but you do have to build the circuit.

In Chapter 2 you learned about the most important tools for this task. This chapter will prepare you for the work to be performed. If you don't have any experience building electronic circuits, we'll give you the necessary information here to make your work easier.

## Buying The Necessary Materials

Each circuit in this book consists of many small, individual parts. They include the electronic components, connecting wires, cables and others. However, before you can work with these components, you have to buy them. While this sounds very simple, it can turn into a problem if you don't pay attention to specifics.

You have already learned about the tools you'll use so we don't need to go into them again here. We'll give you information about other materials here.

### Pay attention

If you plan on buying the materials necessary for the work along with tools, we recommend first reading this book to find out which materials are suitable for the work. Within the individual chapters, we've listed the necessary materials for each circuit in a parts list.

The parts list specifies which components the circuit in a particular chapter requires. Also, you'll need some additional information to make sure you purchase the correct model with some of the components. This information is contained within the text of the chapter in question. So before you run out and buy parts, read the entire text of a chapter to make sure you don't accidentally buy the wrong component.

## A list of the materials

Consider the parts list as a kind of shopping list. These components will be used on the circuit board. In addition, often extra parts will be required. These are elements that you will connect to the board, for example, with cables. Because these elements aren't components that you'll solder to the board, we won't include them in the parts list. Nevertheless, you'll need these components to build the circuit. Therefore, be sure you read the entire chapter before buying the parts. It's too easy to forget a component; especially if you didn't see it in the parts list.

Quite often you will need parts for one circuit that you already used in a circuit from a previous chapter. Depending on whether you want to build both circuits at the same time, you could use these parts for both circuits so that you don't have to buy them twice. This will help save money. The individual chapters also contain information about this.

Before you buy materials, be sure to follow these steps:

1. Check the parts list for the type and amount of required components.

2. Read through the chapter carefully for explanations about the components of the parts list.

3. Supplement the list of components with the additional required components mentioned in the chapters (those which don't get soldered to the board and thus are not included in the parts list).

4. Pay attention to which components you have already used with previous circuits and decide if you will need to purchase these components again or if you can use the ones from the previous circuit.

# Preparing The Materials

After purchasing the materials for a circuit, you will have to modify some of the components before you can use them within the circuit. This is especially true for the resistors.

You'll have to use pliers to bend the wires sticking out from both ends of the resistors by 90°. This lets you fit the wires through the holes on the board and solder them to the back side. Keep two things in mind when you bend the wires:

1. Bend the wires in a small radius, that is, use flat-nosed pliers to grip the wire in the place where you want to bend the wire. Then kink the wire directly on the pliers so that an exact 90° angle occurs.

   You'll have to do the same operation on the other side of the resistor with the second end of the wire. Make certain both bent ends are pointing in the same direction.

2. You have to bend the wires in such a way that the bent ends are at a specific distance from each other. This distance is specified by the grid of the board in conjunction with the component layout of the circuit. The component must fit into the pattern of holes of the board as specified in the drawing. Although this may take some practice, you'll quickly get the hang of it.

## Preparing the wires to be soldered

In the parts lists, we often mention conducting path bridges. These are simple connections between two points of the circuit. Because these connections cannot be made on the foil side of the board due to the circuit layout, they have to take place on the component side.

The conducting path bridges consist of a simple plastic-coated copper wire. We'll describe this wire in more detail later. If you treat the wire according to the following instructions, you'll be able to bend it like the resistors so that the required length of the component is reached.

However, the copper wire we've just been talking about is also used for another purpose. With the help of the wire, the conducting paths are created on the back side of the board. The required connections between the components on the foil sides of the boards are achieved by soldering the copper wire.

Before you run off and buy wire and begin soldering, keep this in mind: There are special wires in the electronics shops which have been specially designed for soldering. We're talking about silver coated wire, in different diameters. For the circuits in this book, you need to get an appropriate coil of wire, with the diameter of approximately 0.025 inches (22ga).

*A coil of silver-plated solder wire*

If you can't find this type of soldering wire, buy some bell wire instead. Don't get the type that is made up of many small wire strands. Instead, get the type that is made of solid copper wire. Also, make certain its diameter is about 0.6 mm and is tin-plated (copper wire that is not tin-plated is poor for soldering). This wire is usually available in small coils, several meters in length.

Before you begin using your coil of copper or silver wire, you will need to do some preparation. First, in the case of the copper wire, you will have to deal with a plastic cover. Also, both the silver and the copper wire are often bent or even kinked from the coil. You cannot use the wire in that condition.

You can change this with a single step. Cut three pieces from the cable, approximately 20 to 40 cm long. Now feed these three pieces into a vice, with the ends of each piece next to each other. If you don't have a vice, use a pliers and work with one piece of wire at a time.

Get another pair of pliers and grip the other ends of the three pieces. Press down hard with the pliers and pull the cables apart from each other (3 to 5 centimeters should be enough). Don't worry if a cable snaps. If you're working with two pairs of pliers and you aren't strong enough to stretch the wire at a length of 40 cm, then shorten the length of the wires.

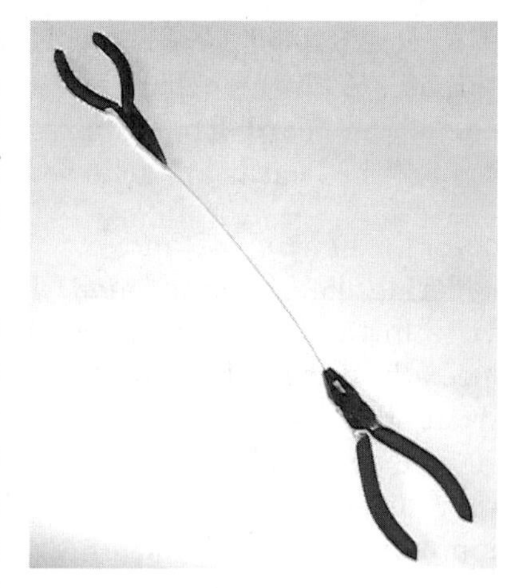

*Stretching the wire*

Stretching the wires has two effects. Since stretching the wire also straightens it, make certain to store it flat. Otherwise, it may bend again. Also, stretching shortens the diameter of the wire. As a result, the cover becomes a bit looser.

After stretching the wire, cut off the ends where you gripped the wires with the pliers so the "crushed" part of the cover is removed. You may also often notice the cover is contracting a bit at the ends, which exposes the wire. If this doesn't happen automatically, help it along by pushing the cover back. Using the pliers to carefully grip the end of the wire and push the cover to the other end. Don't worry if you're not successful the first time. This is often very difficult and requires several attempts. If it becomes too hard, shorten the wires by cutting them into several pieces using the cutting pliers.

If, after you remove the cover, the wire becomes bent again, stretch it until it is straight. Wire prepared in this manner is better for conducting path bridges and for the connections on the back side of the boards.

## Tips On Soldering

After buying the necessary components for a circuit, you have to solder them with the required conducting paths onto the board. Soldering is a job that requires a little coordination and a lot of practice. If you don't have much experience in soldering, you need to get some practice in before you solder your components to the board.

**Important Tip**

If you have no experience at all in soldering, check with your local library, hardware or book store. Many reasonably-priced books on soldering are available that explain the correct procedure.

The following tips will make your work easier.

Always remember when soldering that you're working with heat. You may realize this the hard way if you accidently touch the tip of the soldering iron. Also remember the components can get hot when you are soldering and you could just as easily burn yourself on them.

You're not the only one who is sensitive to excess warmth either. Various components are also heat-sensitive, some more than others. For example, resistors are quite resistant to heat but you must use great caution when soldering in ICs (Integrated Circuits). These parts can easily be damaged if you hold the soldering iron on these sensitive parts for too long or if you use a soldering iron that heats to temperatures that are too high. Therefore, make certain you only keep the soldering iron on the solder joint long enough to liquify the solder and create a connection between the two items to be soldered. As soon as this is done, remove the soldering iron from the solder joint immediately.

After you have removed the soldering iron from the solder joint, the heat continues to spread out over the components. You will easily notice this if you grab onto a component with your hand, for example. Some time may be required for the solder to solidify. However, don't blow cold air on the solder joint to shorten the time period. Blowing on the solder joint can result in a cold solder joint, which can easily break into pieces under mechanical stress. Another effect of a cold joint is that contact between the two components can no longer be guaranteed, resulting in defective circuit functions. Cold solder joints are hard to detect. One indication of a cold solder joint is a dull or unpolished appearance. However, you also get cold solder joints when components are moved while the solder is cooling.

**Important Tip**

If you intend to control high voltage through the terminals, find a nonconducting case and attach it around the circuit. This will protect you from direct contact with the conducting paths. You should also consider this when controlling alternating voltage, even with lower voltage values.

When you are soldering, try to create a solder joint that is as clean as possible. Among other things, this means that you don't use too much solder. Use just enough to connect the component to the solder eye (land, soldering tag) of the board. In the ideal situation, the solder joint will take on a cone-shaped appearance. A ball forms if you use too much solder but the components won't be completely connected to each other if you don't use enough solder, which can easily lead to breaks.

A clean solder joint also means that there is no burned wood rosin at the joint. The flux burns easily and then forms a black wreath around the solder joint. This happens when the soldering procedure takes too long and too much solder is used.

If you have an unclean solder joint, seriously consider removing the solder from the solder joint with an air-suction nozzle and resoldering the area. Since you have to heat up the solder joint with the soldering iron when removing solder, be very careful that you don't destroy the other components with the heat. You will have an easier time getting clean solder joints if you clean the soldering iron tip on a regular basis. Keep a small, damp sponge handy. Then you can clean the soldering iron tip before each soldering operation by wiping off the old solder on the sponge.

Make sure the sponge doesn't get too wet, otherwise the soldering iron could cool off too much when you are cleaning it.

**Important Tip**

When buying solder, make sure that it comes with flux and is designed for electronic applications. DO NOT use acid-core flux solder - it's not designed for electronic projects and will damage components.

## Hints on solder

As you may know, you need the appropriate solder. When purchasing solder, make sure that it comes with *flux*. Flux is used so solder, after it has become liquid, can make contact more easily with the materials you are soldering. The flux is typically a wood rosin and is usually present in the solder. You can tell if flux is present because the solder is hollow in the middle. The flux goes in the hollow spaces. When you solder, this agent quickly becomes fluid and spreads along the solder joint accordingly.

## Conducting paths

For the circuits in Part II of this book, you will have to solder the conducting paths on the back side of the boards with the copper wire we talked about earlier. You have already learned how to prepare the copper wire. Now we'll give you some hints on laying the copper wire.

If you must lay a conducting path, first prepare a sufficiently large piece of copper wire. Make certain it's not too short. While it doesn't have to be the exact length, you don't want it half a meter too long either, since the wire can be very unwieldy.

Begin soldering at one of the end points of the conducting path. As a rule, a component will be connected there with a solder eye. This means there is already a solder joint. Place the copper wire on this solder joint by placing the wire flat on the board so that it is exactly adjacent to the other components. Then heat up the solder joint briefly so the copper wire is connected to the other components.

Next, lay the copper wire as shown in the appropriate drawing. If you follow the drawing, frequently 90°-bends of the copper wire will be necessary. Make the bends exactly as shown in the drawing. When doing this, make sure that the copper wire is always aligned over the solder eye, even after a bend. To make sure that the copper wire rests firmly on the board and cannot be moved, either connect it with all the solder eyes along the way or connect it at regular intervals, for example, every five solder eyes. Although the first method is safer, it does mean more work and more solder. If you opt for the second method, be sure to solder directly before or after each bend. This can also be an advantage in bending. If you put a soldering point right before a required bend, you can then easily bend the copper wire in the new direction using flat-nosed pliers, without having to hold the wire with a second pair of pliers. Make sure that the wire is always flat on the board. Often you will need to heat up a solder joint a second time and press down on the wire with a flat-blade screwdriver.

Follow this procedure from one end of the conducting path connection to the other. Remember, it is not a good idea to add an extra solder joint between two existing solder joints. Since the copper wire expands as it's heated, it can bend to the point where it makes contact with the adjacent wire, thus causing a short circuit.

Finally, we'd like to mention once more that this type of work requires a lot of practice. However, anyone can learn how to do it. So be patient and work carefully to avoid problems.

## How To Take Apart A Circuit

Once you have built a circuit, but no longer have any use for it, you can reuse the components of this circuit to build other circuits. To do this, you'll have to unsolder the components from the existing circuit. When unsoldering the parts, use extreme caution with the heat—you can easily damage parts by overheating them. What we said about soldering parts also applies to unsoldering. Furthermore, when unsoldering parts, be careful not to damage the components by using too much force.

Small hooks can easily develop from cutting wires that are too long with cutting pliers on the rear side of the board. In some cases you bend the components apart, after passing them through the board, on the rear side so that they won't fall off when you solder them. So when you unsolder these parts, they don't automatically fall off, but instead, you have to pull them from the board with pliers.

Check the unsoldered parts for mechanical defects and then clean excess solder from them. Often it is more sensible to spend a dollar for a new part, than to use an old part that is defective. Work with used parts only when you are sure that they are free of defects. You can also reuse the boards, though you'll have to clean them especially well. Since solder eyes (lands or soldering tags) frequently come off when you are unsoldering the conducting paths, we also recommend purchasing a new board in cases of doubt. Often the problems caused by missing solder eyes are considerable and can easily lead to errors.

# Chapter 4:

# The Materials You'll Be Using

# Chapter 4: The Materials You'll Be Using

In this chapter, we'll talk about a few basic electronic components that are used many times in the circuits. We won't go into how these components work. Instead, we'll talk about their special features or characteristics regarding mounting them to the circuit.

Read this chapter to find out what you have to watch out for with the different components.

## The Circuit Board

Most of the circuits we'll talk about are mounted on a circuit board (see illustration below). A circuit board is a fiberglass plate about .04 inches thich (1 mm) and has holes appearing at regular intervals. The holes on the back side of the board are surrounded by copper rings. These rings are used to solder in the components.

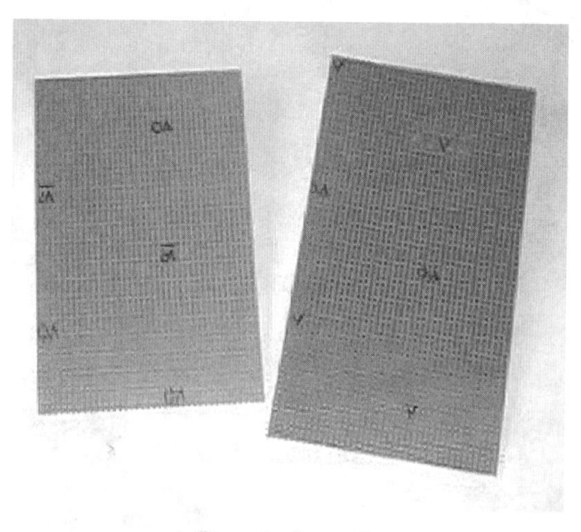

*Circuit boards*

## Using Wire In Soldering

In Chapter 3 you learned about using the required wire. We'd like to mention again that it's best to buy a roll of silver-plated wire of about .24 inches or 22 gauge (0.6 mm). This is the best kind of wire for creating the necessary connections.

## The Resistor

The resistor is one of the electronic components used most often. With its property of resistance to current, it usually makes certain that the other components are not overloaded and thus protects them from being destroyed.

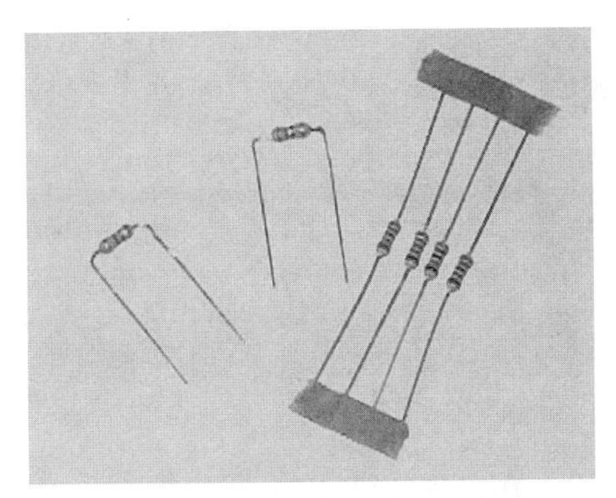

*The Resistors*

The illustration above shows several resistors on the right, similar to the type you could buy from your dealer. The two resistors on the left side have been prepared for mounting in the circuit (in other words, they were bent). The two bent leads are placed through the holes of the circuit board and soldered on the back side.

The resistors have to be able to withstand a load of 0.25 watts (your dealer will probably ask you this question when you buy resistors).

## Diodes

Many circuits use *diodes*. We'll talk about two types of diodes in this section.

### The "normal" diode

The "normal" function of a diode is to allow the current to pass in one direction and to block it in the opposite direction. For this purpose, the diodes, like the resistors, are soldered into the circuit on the circuit board.

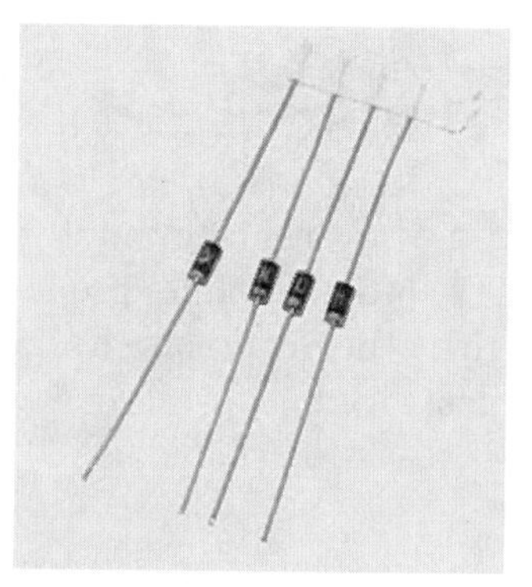

*Normal diodes*

For the diodes to be able to perform their function, you need to mount them properly in the circuit. This means you must consider the polarity of the component. In the figure you see a gray ring around the case at the bottom borders of the components. This is how the cathode, or the minus pole, of the component is marked.

You will find the symbol on the right used for diodes in the drawings for the circuits. The triangle in the symbol is pointing at a line that is perpendicular to it. This line represents the minus side, that is, the cathode, in the plans. In building the circuits, keep to the direction in which the components are to be mounted, to avoid defective functions.

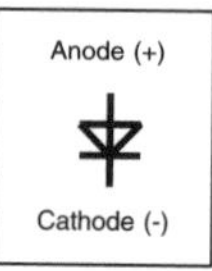


### Light-emitting diodes (LEDs)

LEDs are a special type of diodes. They aren't used for the same reason as the "normal" diodes we just mentioned. The purpose of an LED is to emit light, hence the name "light-emitting diode" or LED.

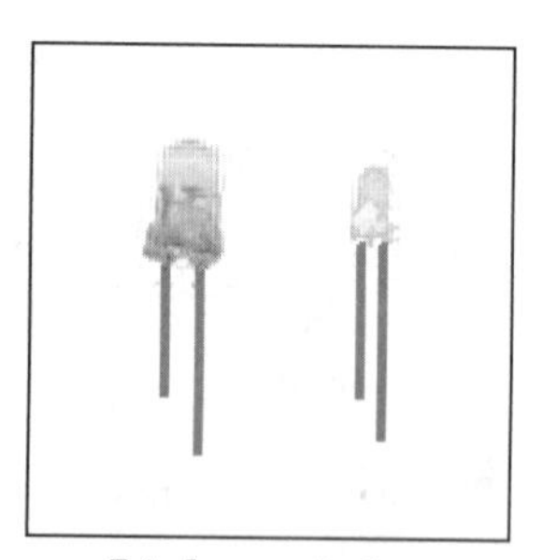

*Light-emitting Diodes (LEDs)*

In the instructions, the light-emitting diodes are represented by the same symbol. You'll recognize the anode, which is the plus pole, and the cathode on this component by the length of the connecting wire. The longer connecting wire is always the plus pole, or the anode. That means that the cathode is the shorter connecting wire.

After using a light-emitting diode once, and having cut off the connecting wires accordingly, in most cases you will still be able to detect the cathode by another attribute. The plastic coating of the light-emitting diode is often flattened on one side. The connection attached there is the cathode. This way you can use the diode correctly a second time.

## Capacitors

Some circuits require capacitors. There are various designs of these components. The following illustration shows three models of capacitors:

The capacitor on the left is called an electrolytic capacitor. It has both a plus contact (longer connecting wire) and a minus contact. The minus and plus contacts are marked accordingly on the case. The capacitor in the middle is called a tantalum capacitor. It also has a polarity. The longer connecting wire is again the plus pole. The component on the right is called a foil capacitor. It has no polarity and can be mounted without regard to direction.

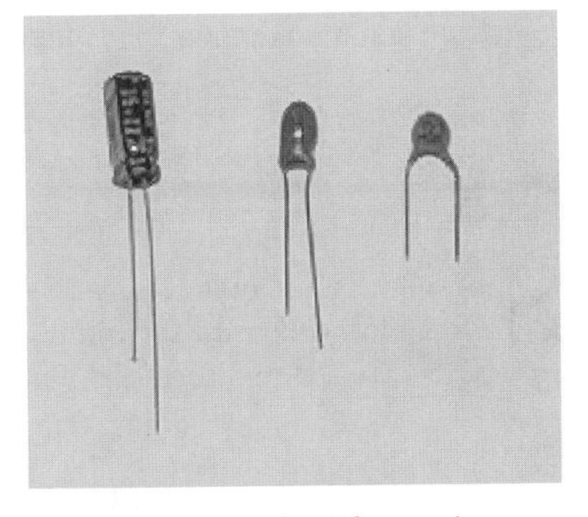

*Three examples of capacitors*

The parts lists for the individual circuits specify which types of capacitors to use.

## IC Chips And IC Sockets

IC chips are integrated circuits that connect to the circuit using several pins. Since these components are quite sensitive, they are frequently mounted in combination with an IC socket. The IC socket is soldered to the circuit.

Then you plug the IC chip into the socket, thus making contact to the circuit. In this way, you can easily mount this sensitive chip, and when necessary, quickly pry it out of the socket, for example, to use the IC chip in a different circuit.

*IC-chips and IC sockets*

When you mount an IC socket, make sure you insert the socket with the marked side in the correct direction. The marking is sketched in the component plan. Also, when you mount the socket, make sure to use an IC socket with the correct number of pins. When you plug the IC into the socket, match up the marking on the component with the marking on the socket.

## Plugs

Some of the circuits are add-on boards that you'll then plug into a motherboard that you created earlier. To make it easy to plug the add-on boards into the motherboard, you will have to buy 21-pin strips.

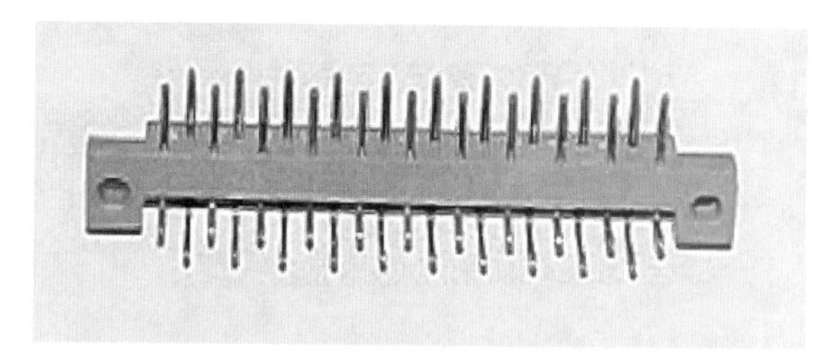

*A 21-pin pin strip*

The pin strip is angular so the rows of pins will fit into the pattern of the circuit board. You need the appropriate counterpart, a socket strip for the motherboard. Along with these strip combinations, many circuits also use edge connectors.

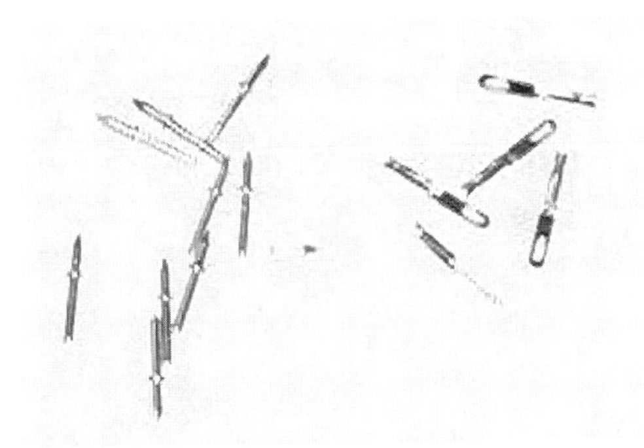

*Edge connectors*

The male connectors come in different strengths. The ones with a diameter of 1 mm will fit well in the circuit board. The male connectors with 1.33 mm diameter often create problems; you can only plug them in by using a lot of force or by first widening the holes. This often completely removes the solder eye on the back side. Male edge connectors are available in matching sizes to the appropriate female connectors. Here again, pay attention to the diameter.

## Relays

Some circuits use *relays* to act as controllers on add-on elements connected to the circuits or to give you the option of using the closing contacts for your own purposes.

The relays we'll use in this book are usually miniature relays that can be operated with 5 volts. However, these small relays can only switch small currents or voltages through their contacts. Be careful when using these small relays. For example, they aren't always designed for 220 volts. There are also more powerful relays, which are somewhat larger. These relays also have to be activated with 5 volts, but the power to be switched can be set higher.

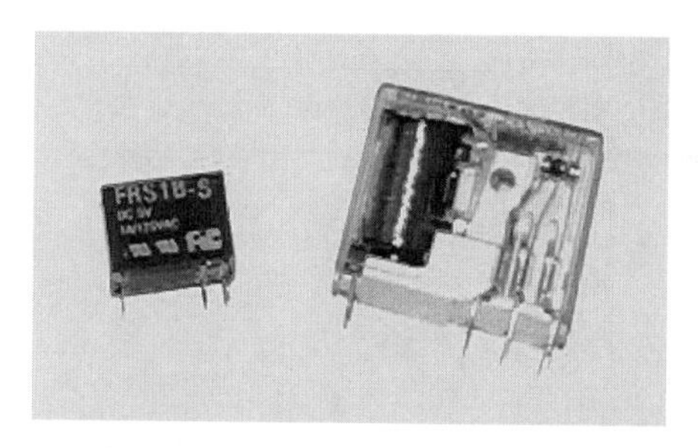

*Miniature and power relays*

As a rule, the relays have the same contact layout. When buying the relays, make sure that they will fit on the contacts shown in the component drawings.

## Voltage Regulator

The last component we're going to talk about in this chapter is the voltage regulator. This component's task is to guarantee the required voltage of 5 volts on a continuous basis. For this purpose, it is necessary that a somewhat higher input voltage be applied to the component. The component then adjusts this voltage down to 5 volts.

The three pins of the regulator have the following functions:

1. When you look at the component in such a way that you can read the label, the left pin receives the higher input voltage. It serves as the input.

2. The middle pin is the ground cover, both for the higher input voltage and the 5 volt output voltage. This pin is connected to the metal plate on the rear side of the component. This metal plate carries off the resulting heat using an external cooling body or heat sink.

3. The right pin provides the 5 volt output voltage. The required voltage for the circuit is picked up here.

Along with the components listed here, additional electronic components will also be required at times; we'll discuss and describe them in the appropriate chapters.

# Chapter 5: Serial And Parallel Ports

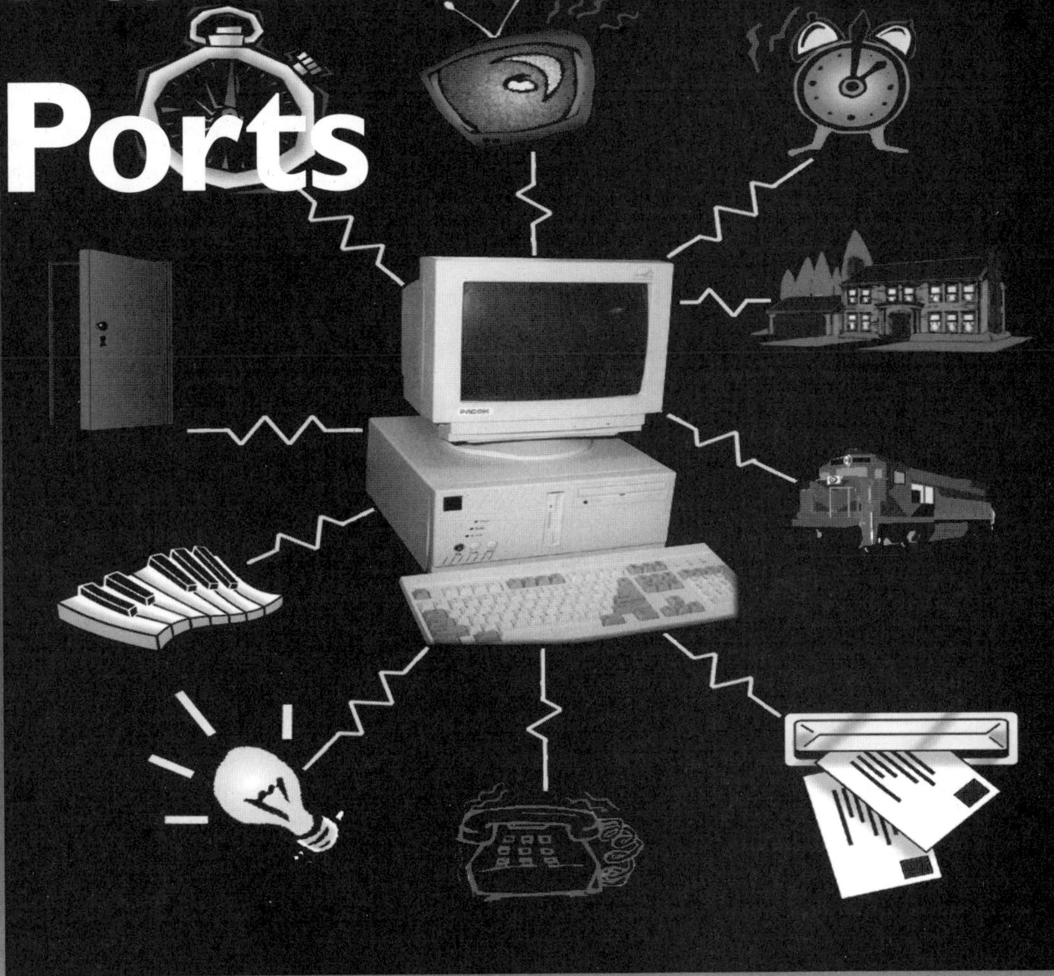

# Chapter 5: Serial And Parallel Ports

Exchanging data between your computer and the circuits we'll talk about in this book will occur through the ports of your computer. We'll talk about two types of ports in this chapter: the serial port and the parallel port. You'll learn how these ports work and how they differ from one another.

## Serial Port

You've probably already used the serial port on your PC more than once, perhaps without even being aware of it. For example, this is where you connect the mouse on most PCs. You will easily recognize the serial port by a 9-pin connector on the rear of your PC. However, a 25-pin connector is also being used for the serial port now.

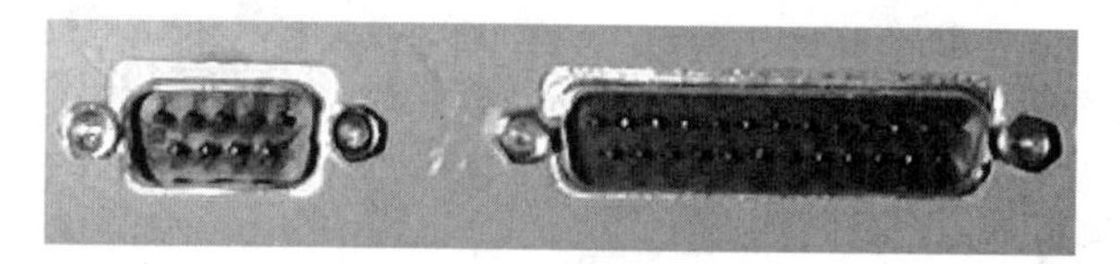

*The serial connectors on the PC*

In many cases, today's computers include two serial ports; frequently both connectors are used. Keep in mind that you're using male connectors with both sizes. In other words, you will see pins (unlike the connector of the parallel port, which is also 25-pin, however, it has pin holes because it is female). The serial ports are the communications ports of your PC and are often simply called COM1 and COM2. Any terminal or modem program can access these ports and use them for data transmission. The serial port is also used to control some of the circuits we'll describe in this book.

Just what is a serial port? Probably the most crucial feature of this connection lies in the word "serial." Serial means "one after the other" and this is exactly how data is sent to the serial port through the connection cable. In general, each character in a PC consists of one byte (one byte is the same as 8 bits). In a transmission through the serial ports, the 8 bits are sent to the cable one after the other like pearls on a string. A device or a program that expects the data in this precise form must be at the other end of the line. This device or program puts the 8 bits back together to form one byte, that is, a whole character. To ensure that everything runs smoothly, additional bits are added to the actual data for checksums and other information. In view of this simple method, which contains many steps and additional data, you can imagine that high transmission speeds cannot be expected in serial transmissions.

Although serial ports are slow, they do have some important advantages:

- Serial ports exist in almost every type of computer, from a mainframe with several CPUs or computer network solutions to IBM compatible PCs to programmable calculators, electronic organizers or clocks. Thus, even the most different operating systems and programs can exchange data relatively easily.

- The basic definition of data processing and pin layout has been available for many years. In 1969 the Electronic Industries Association (EIA) established the RS-232 Standard, which has remained unchanged ever since.

  The basic cabling of the RS-232 definition manages with a 3-conductor cable for the connection of two computers. One conductor is the ground wire and the other two conductors are used for transmitting the data. One of these two conductors sends the data and the other conductor receives the data. The simple cable is inexpensive, already exists in most offices as telephone connections and is also well-suited for a connection over great distances. In principle, a connection using modem to another computer with a modem halfway around the world is nothing more than a three wire connection of two computers in the office. The only difference is that the distance is much greater.

- If a data transmission doesn't have to be high speed, the serial port with its progressive chip technology and simple concept is ideal. This is why the first "pointing devices", such as the mouse or drawing pads, were connected to a COM port. Such devices don't handle huge amounts of data. Usually, information is only sent sporadically and in irregular amounts.
- Many machines also have serial ports. The first computers had the complicated task of controlling machines and factories. The computer analyzed information from a sensor and then, if necessary, changed the work routine in the machine itself. It received the information from the sensor through the first serial port and conducted the control of the machine through the second port. Some of the circuits in this book function according to a similar principle.

You're probably wondering why 9-pin or 25-pin plug connections are used for a three-conductor connection? What about the other pins? You don't need to look too hard for the reason: The EIA and another standardization committee, the CCITT, have reserved an additional 17 other lines with signals. These signals are used to control the information flow, to make things more secure and flexible. Fortunately, no one is forced to use or evaluate all these signals. With a 9-pin plug, that wouldn't be practical anyway. Only three conductors are really necessary to make serial transmission possible in two directions.

With a connection to a modem, there are six additional signals, used for better communication between the modem and the PC. The PC and modem use these signals to control the flow of data. For example, this is necessary when the modem receives data much more quickly than the PC is able to process the data. The signal lines are necessary to slow down the modem. If this weren't the case, data would be lost, because the receiver could not pick up the data quickly enough.

We'll explain the lines we've been talking about using the pin layout of a 25-pin connector. Here's what it looks like:

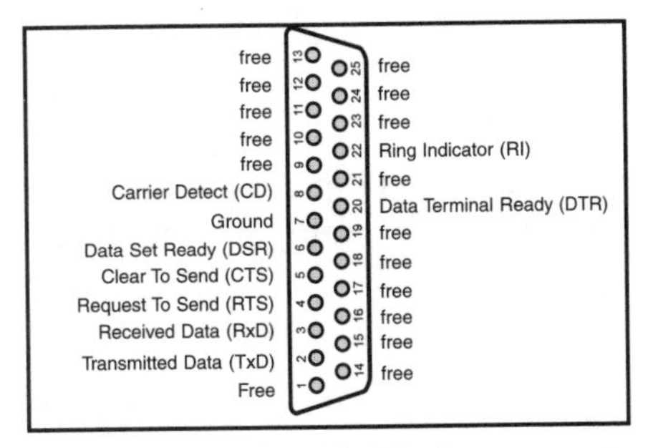


*Diagram of a serial 25-pin connector*

## Pin 2

Transmitted Data (TxD) The data is sent over this line.

## Pin 3

Received Data (RxD) means that the data is read here. All data arrives here serially over this pin.

## Pin 4

Request To Send (RTS) is used by the PC to ask the modem whether it is ready to receive data.

## Pin 5

As soon as the modem receives an RTS signal and is not currently busy with any other tasks, it sets the Clear To Send-Signal (CTS). This lets the PC know that it can send.

## Pin 6

The modem sets the Data Set Ready (DSR) signal to let the PC know that the connection to the other party has been successfully made.

## Pin 7

The Ground line ensures that the individual signals can receive a common voltage, namely 0 volts.

## Pin 8

Before the DSR signal is set, indicating a successful connection, the Carrier Detect signal (CD) is set. The modem uses this signal to communicate that there is someone on the other end of the line who is apparently able to do something with the signal.

## Pin 22

The modem sets the *Ring Indicator (RI)* as soon as it receives a ring signal; in other words, as soon as someone calls the modem. For example, a software program uses this signal to regulate the number of rings necessary before the modem receives the call.

As you can easily see, in a direct wiring between two computers, there are a few signals to be synchronized. For example, Pin 2 is wired with Pin 3, so that the outgoing signal on the one end arrives at the incoming signal pin at the other end and vice versa.

The following figure shows the pin layout of a 9-pin connector:

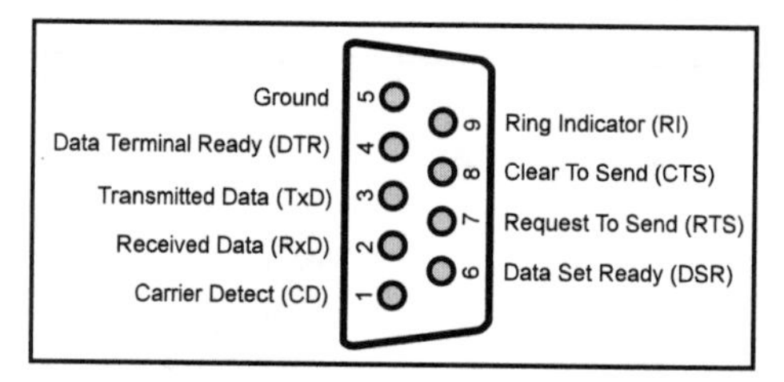


*The pin layout of a 9-pin connector*

So, with some lines the PC waits for signals from an external device, while at other lines it sets signals for these devices. The actual data is only sent over the two data lines. This process is ideal for controlling all machines that are dependent on simple signals and do not rely on high speeds. This is precisely the reason why some of the circuits in this book use the serial ports of the PC. Terms that we explain here will be encountered frequently in the book.

## Parallel Port

One of the first differences you'll notice between a serial and parallel port (abbreviated LPT) is that they look different. Although the parallel port also has a 25-pin connection, the connection is female. So, a cable that plugs into the parallel port must have a male connection.

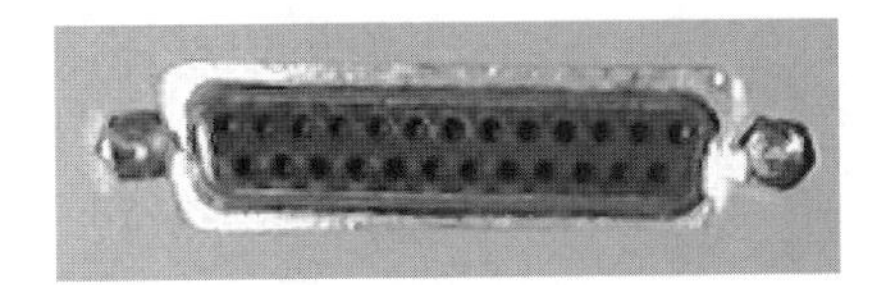

*The female connector of the parallel port*

However, the main difference is the parallel transmission of data over 8 data lines. In this way, 8 bits (1 byte or 1 character) can be transmitted at the same time. This corresponds to about eight times the transmission speed of a serial port.

The parallel port is normally used to control a printer. Although it was once possible to use a serial printer, today's demand for faster printers and the power of today's printer languages (such as PostScript) require using the faster parallel port.

The parallel port, as a result of its allocation for controlling a printer, is only meant for data transmission in one direction. After all, the printer receives data. The parallel port doesn't have a data buffer for receiving data like the serial port. However, besides the eight data lines, the parallel port also uses signal lines to communicate information to the port from the printer. For example, there are signal lines that let the PC know when the printer runs out of paper or when it is "offline." So if we send data to the parallel port, we can "misuse" these signal lines for the transmission of data. In this way, some of the circuits of this book will send their data over the parallel port to the computer.

If you look closely at a printer cable, you'll notice the one end of the cable has a plug that fits the 25-pin connector. At the other end, there is a different plug. It is called a Centronics plug and contains 36 connections. The specification of a printer port, i.e., the connections, plug and signal assignments, are equated with the term Centronics. The shape of the plug, the signal assignment and even the properties of the cable are accurately stored there.

The following figures describe which signals go with which pin of the 25-pin parallel connector or of the Centronics plug:

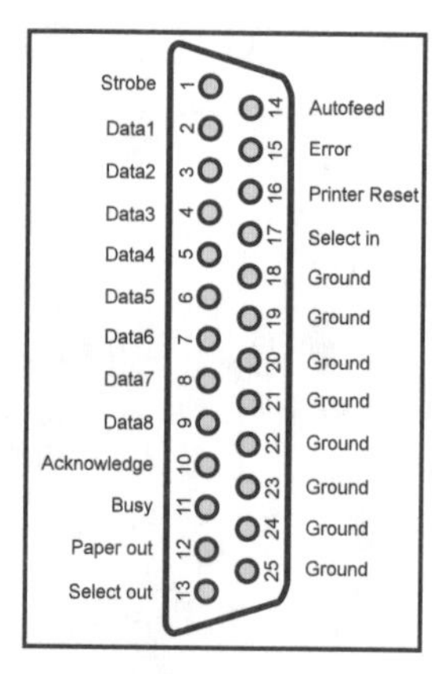


*The connector of the parallel port (female)*

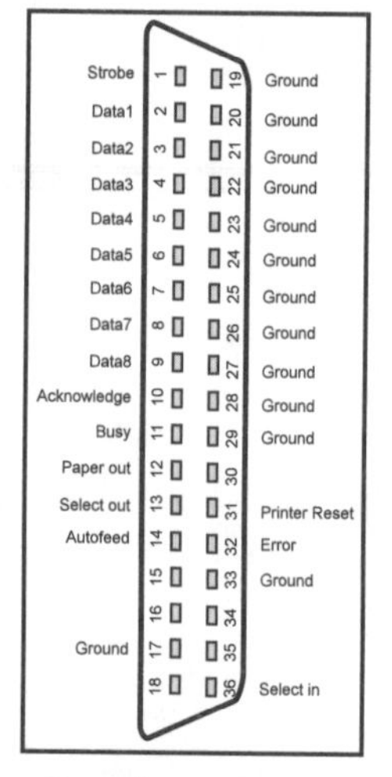


*The Centronics plug of the printer cable*

Your PC won't care whether the parallel port receives signals from a printer or signals and data from a machine or a circuit. What is important is that the PC has a program running that recognizes the signals of the circuit and evaluates them accordingly. For this purpose, we have included a program that performs this very task for each circuit of this book that uses the parallel port.

# Chapter 6:

# The Software For The Circuits

# Chapter 6: The Software For The Circuits

Before we start Part II, where we describe the circuits you'll be creating, we need to talk about the software we'll use to control the circuits.

## Why Do You Need Software?

The circuits in this book are designed to work with your PC. The circuit performs the electronic tasks that are created by correctly assembling and wiring the components. The software is responsible for controlling the PC. The PC must react in the manner required for the functioning of the circuit. How the PC behaves and which tasks the software has to perform varies from circuit to circuit.

One of the software's tasks is to interpret user input. Your input influences the behavior of the circuit. You pass information to the software using the mouse or keyboard and the software sends the information through the ports to the circuits, which brings us to another important task that the software has to perform.

This task consists of preparing the information for the circuit so that it can be communicated over the ports and the circuit can process the information properly. Another task for the software is addressing the computer ports for smooth exchange of data. This is absolutely necessary, regardless of whether data is transmitted from the computer to the circuit or vice versa.

Furthermore, the software is responsible for gathering data, which is then transmitted to the computer, the software must read and interpret the data for you so that, for example, the result will be visible on your monitor. If necessary, the software analyzes the data and responds accordingly. For example, the circuit could gather the data and transmit it to the software through the port. The software then analyzes the data. On the basis of the analysis, information might be transmitted back to the circuit. This is called a *feedback control system*.

As you can see, the circuits will not function without software. The software is responsible for important functions; it's not possible to use the circuits without it.

## Using Visual Basic To Create The Circuits

We used Visual Basic 4 (VB4) to create the software for the circuits in this book. VB4 Professional contains both a 16-bit version for Windows 3.1x and a 32-bit version for Windows 95. We created it using the 16-bit version. In programming the software, it is necessary to create *source code files*. These files contain the information and commands that the software is ultimately supposed to execute. We wrote a separate software program for each circuit, meaning that we also had to create corresponding source code files for each software program.

The companion CD-ROM includes these source code files (among other files). Each circuit has its own directory on the CD-ROM. The subdirectories are labeled according to the chapter and project numbers. Each project's directory has all the data that is needed for the circuit. Each project directory also has a subdirectory called *Source*. The source code files for the software of the project are located in this subdirectory. If you're planning to modify the software, for example, to adapt it to your own special requirements or to expand the software's functions, you can use these source code files as your foundation. Simply load the source code files in Visual Basic 4 (16-bit version) and then make the desired changes.

**Important Tip**

Remember that most source code files use a special DLL (Dynamic Link Library) to address the ports. This special DLL is also located on the CD-ROM. It is located in the Windows subdirectory, and is called Portio.dll for the 16-bit version. If you're planning to adapt the software to the 32-bit version, you'll need a file called Portio32.dll. This is the 32-bit version of this DLL and is not on the companion CD-ROM.

Don't panic...you won't have to learn any programming to use the circuits. We give you the necessary software already compiled and ready to use for each circuit.

# CD-ROM Installation

This section introduces the software on the CD-ROM. The directories are titled by chapter. For example, Chapter 8 equals chap_08 on the CD, Chapter 9 equals chap_09, etc. There are two ways of copying and installing the programs. You may run the SETUP.EXE program located on the root of the CD-ROM. This will install all the project's programs including the source code into a directory on your hard drive. Or, you can simply copy the projects one at a time using Windows Explorer.

You may also install the shareware program PC TRACE located in the PCTRACE directory. Click **Start**, select **Run...** and type D:\PCTRACE\SETUP.EXE. The program will be installed on your PC. PC Trace was written by Ehlers Technical Consultants, 4520 S. 58th St., Lincoln, NE 68516. This product is shareware and if you use the programs, you're required to buy a license to continue using them. Be sure to read the README.WRI to comply with the agreement.

You'll find several files in each directory. They include the .EXE (executable) and source files (.FRM, .BAS and .VBP). The dynamic link library is also included in each folder (portio.dll).

## Running SETUP.EXE

Click **Start**, select **Run...** and type D:\SETUP.EXE. You'll see the following screen.

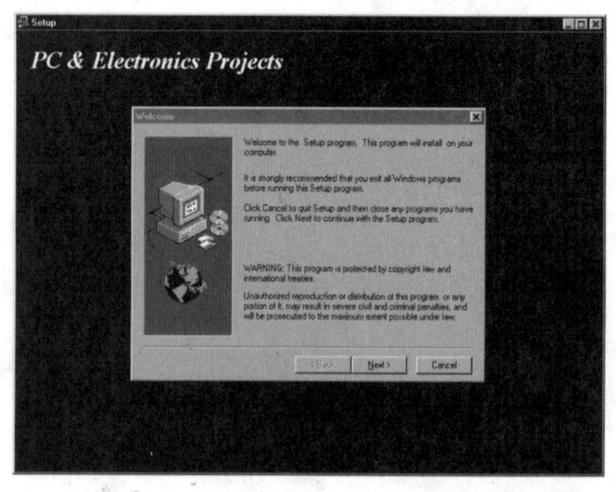


*The Welcome Screen soon appears...click* Next > *to continue.*

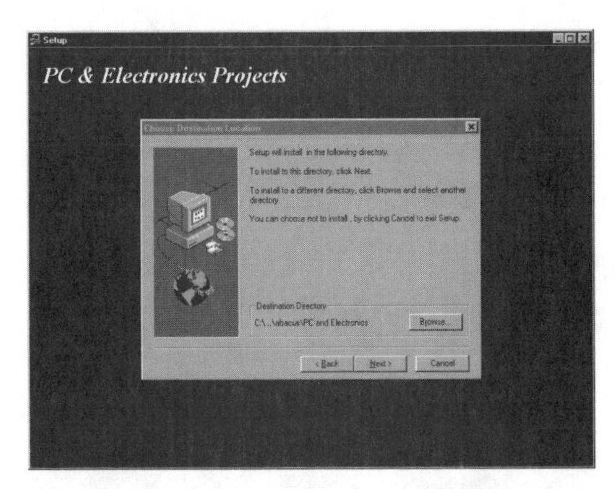


*Select your destination and click* Next > *to continue.*

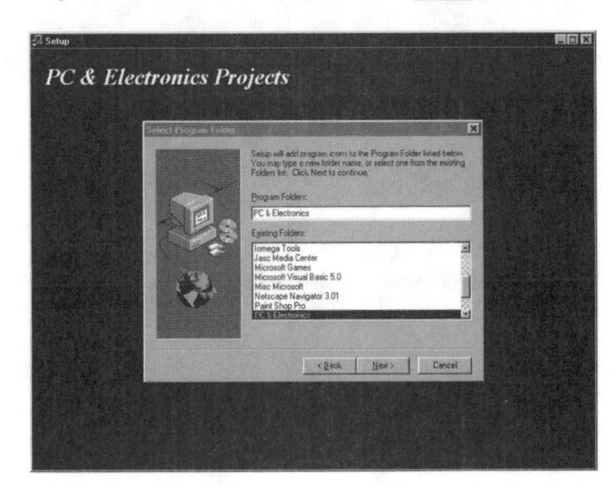


*Select your folder and click* Next > *to continue*

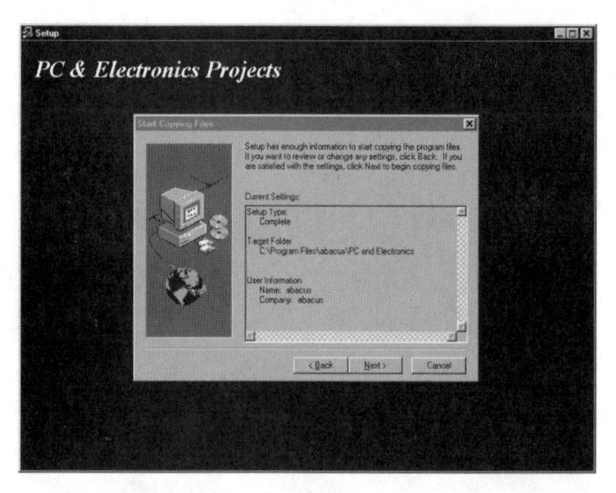


*Review the settings and click* Next > *to continue*

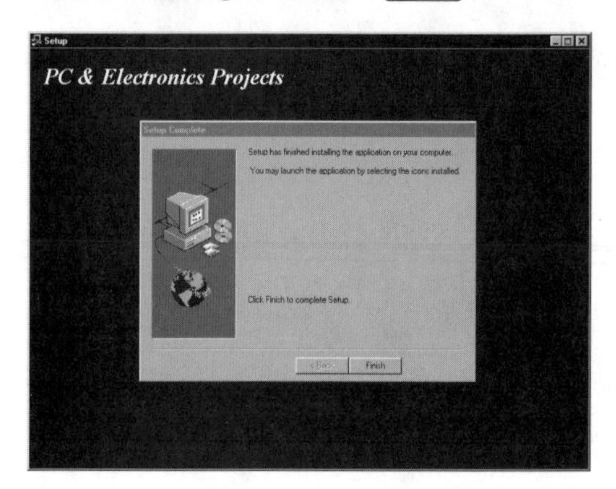


*The installation is completed and click* Finish

*The program window.*

After the installation you can load and run all the project programs. Simply click **Start**, select **Run...** and locate the project you want to run.

# Part II: The Circuits

Part II describes the circuits that you can build and use with your PC. Each chapter describes a separate circuit with all the information you need to build the circuits yourself. The first circuits are simple so even beginners can become familiar with the tasks and necessary steps.

Although the circuits eventually get more difficult, you'll quickly master them if you built the ones described before them. Also, review the information from Part I. It has many helpful hints to make the practical work easier. If you don't build all the circuits, or if you build them in a different order than we use, it's possible that you'll need something in a chapter that was explained or built in an earlier chapter. If so, read over the preceding chapter(s) to get the information. Even if you don't build the circuit, the information in the chapter may help you in your work with other circuits in the book.

In any event, be sure to build the first three circuits; they'll help you become familiar with the procedure used in this book. Chapters 7, 8 and 9 contain basic information that applies to the circuits described in the following chapters. Chapter 9 also introduces a motherboard that will be used in combination with several circuits.

# Chapter 7:

# Serial Port Tester

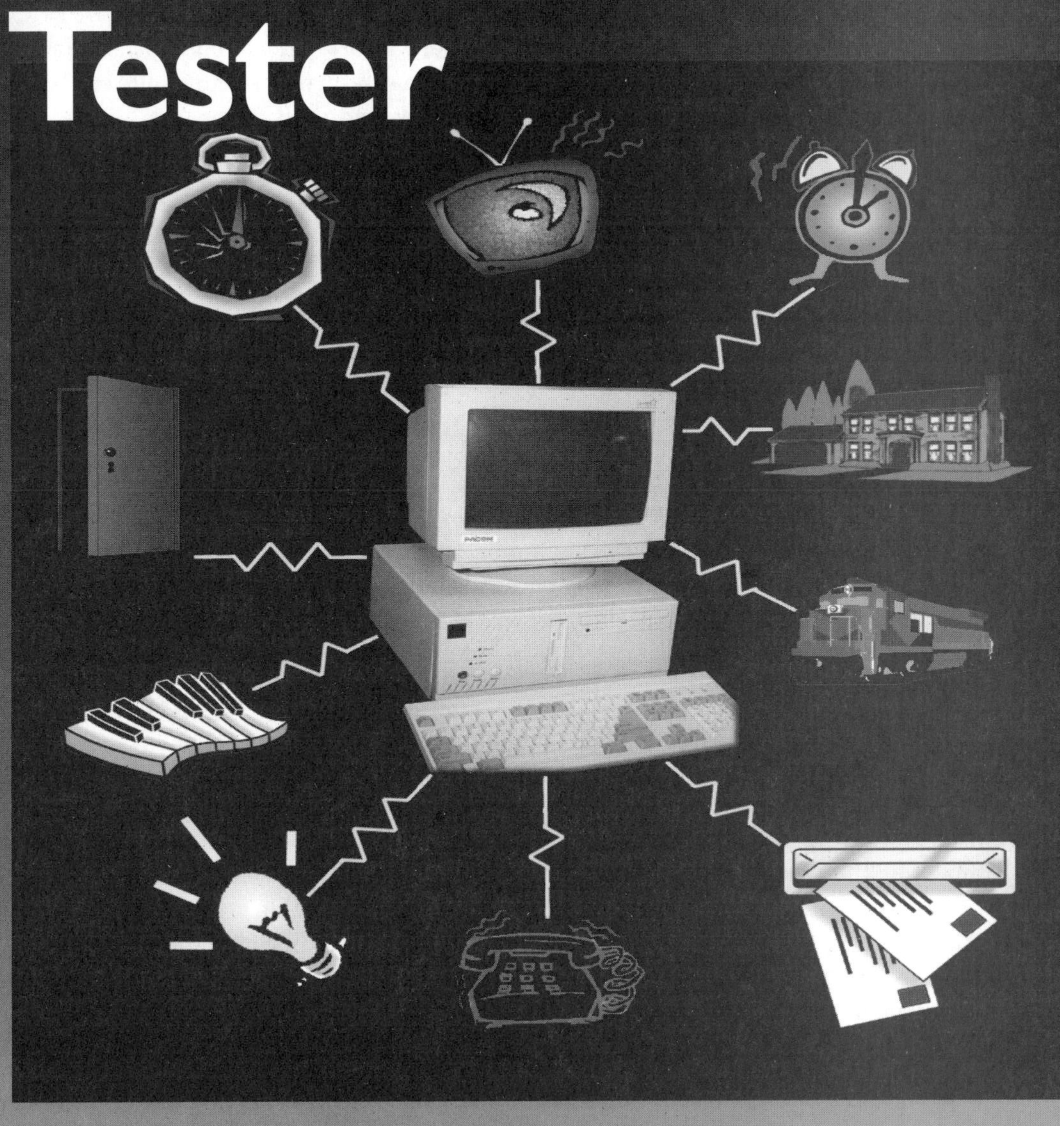

# Chapter 7: Serial Port Tester

In Part I you learned a little about how the serial port works. Now, we'll build a circuit that uses the serial port. Since you can build this circuit with just a few components, this is an ideal "starter" circuit and will help familiarize you with the necessary soldering work and procedures.

## The Circuit And Its Function

The idea behind this circuit is to show you the theoretical information you learned about the serial port in Part I does apply. This circuit will show how data is transmitted over the port in conjunction with signals.

Since the signals usually cannot be detected, the circuit is designed to make the signals visible. This occurs through light-emitting diodes that either shine or do not shine at specific signal states of the corresponding port lines. Since the signals frequently change their states during a data transfer, the light-emitting diodes let you recognize information is actually being exchanged there. Thus, the idea behind this circuit is not to test the serial port to see whether it is functioning correctly.

Now that would be going too far. However, you can use this circuit to test whether the circuit is exchanging data. So, at the same time, you're conducting a basic test for the other circuits in this part of the book that also work with the serial port.

The circuit is designed so that you plug one side of it into the serial port of your computer while connecting a device to the other side, e.g., a modem, without impairing the function. You can insert the circuit between a serial end device and the serial port without function restrictions. This is possible because the signal and data lines run through the entire circuit without interruption. Since the signals can change their states very quickly, the function of the port is frequently recognizable only by the flickering of the light-emitting diodes. Individual circuit states are not perceptible during a data transfer.However, before you can experiment with all of this, you'll have to build the circuit.

## Building The Circuit

As you have already learned, the circuit is built on half of a European circuit board. The European circuit board is 4 inches wide (100 millimeters) and 6.5 inches high (160 millimeters). The board for this circuit requires a height of 3 inches (80 millimeters). We'll have to cut the board in half (for a size of about 4x3 inches).

**Important Tip**

You'll read references to this circuit board throughout this book. You can buy a "European" circuit board at most Radio Shack stores. These blank circuit boards are also called "proto-boards" or "pre-perfed boards." These are available with or without foil on one side. You can substitute the author's European board with proto or preperfed boards.

### Preparing the circuit board

You could use a saw to cut the board in half. But if you do, you'll have to use a fine saw and it will require a lot of time and create a great deal of fine sawdust.

Therefore, we recommend using a utility knife. You'll find that cutting the board is much simpler and faster using a utility knife. Mark the row of holes where you wish to cut the board. Make sure you do not mark the side with the solder eyes. Next, place a ruler along the line marking the row of holes and move the utility knife along the ruler so the areas between the holes are scored. Then deepen the cut line by moving the utility knife along the ruler a few more times.

Once the spaces between the holes have been scored, find a flat surface with an edge (a desk or table is good). Place the board with the solder eyes face down on this surface so one half of the board lies flat on the desk and the other half extends beyond the flat surface. Make certain the carved line lies perfect along the edge of the desk. Now take one hand and press on the half of the board that is on the desk. You'll want to do this so the board won't slip. Then use your other hand to press down on the other half of the board so it snaps off. Be very careful and increase pressure gradually. The board will break in two quite suddenly. To prevent injury avoid letting your hand slip.

Once the board is cut, the copper solder eyes frequently remain intact and are merely bent. To cut the solder eyes, bend both halves of the circuit board back and forth a few times. The solder eyes will soon break off. If you wish, use a file to smooth the area where the break occurred. However, this isn't always necessary because the edge won't be sharp enough to cause severe injuries.

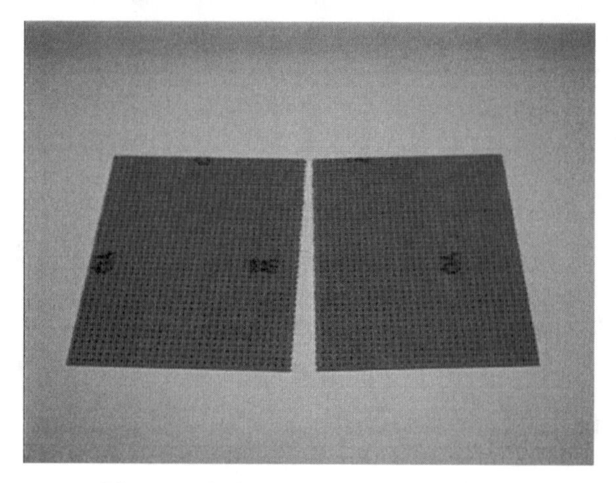

*The two halves of the circuit board*

## The elements on the component side of the circuit board

After cutting the circuit board in half, set one half aside and save it for another circuit. You'll use the other half for the circuit described in this chapter.

First, let's turn our attention to the component side of the board (the side without the solder eyes). This is where we'll place the electronic components. Most components have pins that go through the holes of the circuit board. On the solder-eye side of the board, these pins are soldered to the solder eyes and connected to each other with small wires.

The port circuit contains several components that must be precisely placed on the board. There are seven light-emitting diodes and seven resistors. You cannot arrange and solder these components randomly because the circuit might not work properly then. To help you determine the correct layout of the components, each chapter has a drawing to which you can refer to find the correct position of the components. The solder eyes are displayed in outline so that you can count them and transfer the drawing to your circuit board. The following figure shows you where to place the components.

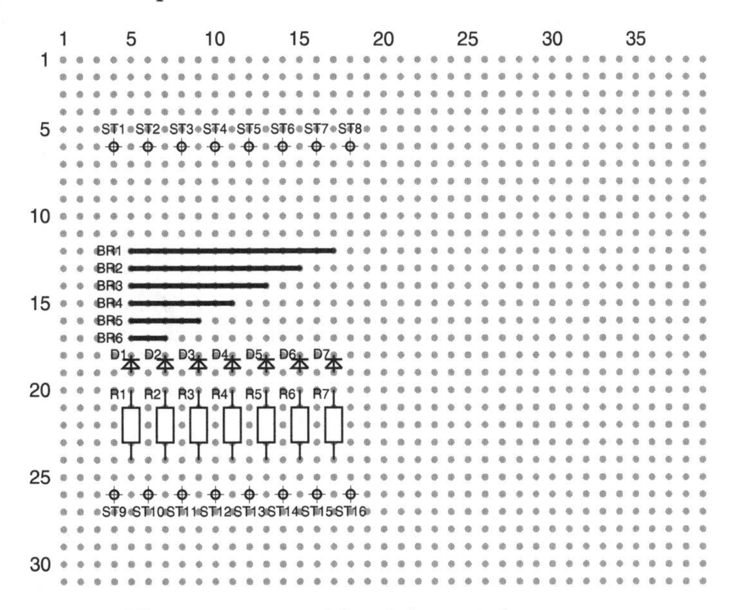


*The component side of the serial port test*

There are numbers next to the rows of holes to help you count and find your way around on the circuit board. Use the numbers to determine precisely where to insert the appropriate components. The components are identified by letter/number combinations. However, this label doesn't tell you with which component you're dealing.

**Important Tip**

As you can tell from the drawing, all the components are placed on one side of the board. The other side remains empty. You are free to divide this board into halves again. Simply use the same procedure we talked about earlier.

To ensure that you mount the correct component in the correct place, we have included a "Parts List" for each circuit plan. The parts list includes the labels for the components, which are assigned to the corresponding electronic components. In this way, you can immediately recognize which component goes with which label.

## Parts list

The following parts list contains the components for the circuit drawing on the last page.

| Parts List | | |
|---|---|---|
| Label | Name | Component type |
| BR1 - BR6 | Conducting path bridge | Copper wire |
| R1 - R7 | 1 kΩ | Resistor |
| D1 - D7 | Light-emitting diode, 3 mm, red | Light-emitting diode |
| ST1 - ST16 | Connectors | Pin connection |

You'll find the resistors and light-emitting diodes here. Using the labels, you can locate the individual elements on the board and build the circuit.

## Building the circuit

Building a circuit is generally easy but it does depend on its complexity. However, let's not forget the old cliche: Practice makes perfect. If you're building a circuit for the first time, you'll have trouble with the soldering and some other things. The more work you do, the less trouble you'll have.

When building a circuit, always pay attention to what you're doing. Each component type has its own peculiarities. For example, with the circuit in this chapter you have to pay attention to the polarity of the light-emitting diodes. Since light-emitting diodes have an anode and a cathode (that is, both a plus and a minus connector) you have to mount these components into the circuit accordingly. If this isn't considered, you risk building a circuit that won't function. In some cases, this could even result in destroying that component and other components.

In addition to position specifications, the component plan also contains additional information about the components. You'll use this information to mount the components properly. Often it's a good idea to refer to Part I where it introduces different components and describes their special features. For example, the component drawing of this circuit contains information about where the positive terminals of the diodes have to be mounted.

If you look at a light-emitting diode, you'll notice that the positive terminal is not identified with a plus sign. The positive terminal is labeled differently on a light-emitting diode. Note that the two pins of a light-emitting diode differ in length. This is how the positive terminal is labeled: The longer of the two legs is the positive terminal.

**Important Tip**

Often identical components have different types of labels. For example, some round light-emitting diodes also have a flattened side. This identifies the negative terminal of the diode. Pay careful attention to labels, and be sure to read the manufacturer's or distributor's documentation.

Resistors have no polarity so it doesn't matter how you mount the components into the circuit.

With the components we've discussed, you'll also find the terms ST1 - ST16. These are edge connectors. They are used to make contact to other components through the appropriate plugs, which, in turn, are soldered onto wires. We'll tell you more about them later.

1. Once you have all the information you need, you can build the circuit board. To do this, place the first component in the proper position on the board, turn the board over and solder the component leads to the appropriate solder eyes.

2. Use a pair of cutting pliers to snip off the excess component leads, above the soldering point. Be careful not to cut off any solder. Cut the lead where there is no solder.

3. Do the same with the other components, until they are all attached to the board.

*The components on the circuit board*

## The Foil Side Of The Board

After you have attached all the components, turn the board over. You'll find the soldered joints there. Naturally, the necessary connections between the individual components are still missing. You'll create the connections using small wires that you lay from one solder eye to the next. It's also possible that a wire is soldered to an existing connection and then has to be dragged to a solder eye or to another wire. The wires are always aligned at the pattern of holes, so that when the wires are laid, only 90° angles are set.

The necessary conducting path connections for the circuit are also depicted for you in the drawings. These describe the wires to be laid, and you must also follow the position for these as accurately as possible. This drawing also contains numbers at the border, which make it easier for you to get your bearings. Keep in mind that we count from top to bottom in this drawing, but from right to left.

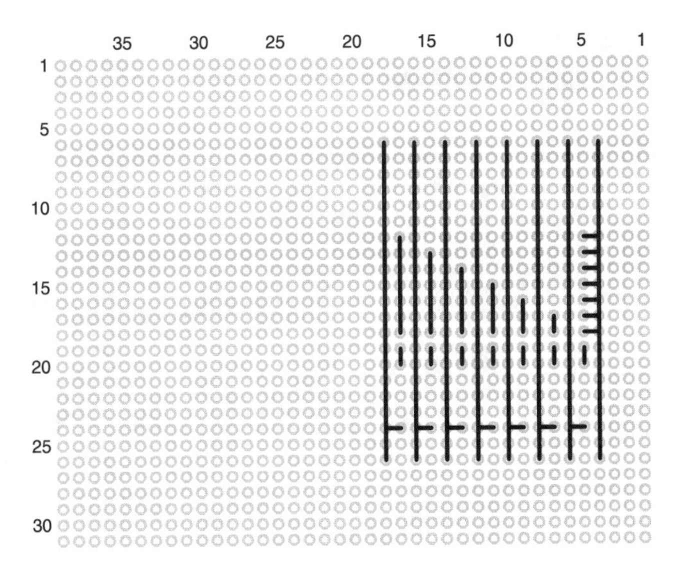


*The foil side of the circuit*

Now you replicate the required connections from the drawing to the board.

1. Find a connection that you would like to make. Take the prepared wire (see Part I) and solder it to the corresponding soldering point. Lay the wire according to the drawing, piece by piece.

2. Always solder the wire to a solder eye: either every five to ten solder eyes or as soon as you need to make a bend in the wire. Then solder the wire to the solder eye before the bend. This fixes the wire in its position and lets you bend it easily in the proper direction with flat-nosed pliers. Use a screwdriver to press down the wire whenever you're setting a soldering point, so that the wire lays flat on the board.

**Important Tip**

You could also solder the wire to each solder eye. In this way, your conducting path will be attached firmly to the board. The disadvantage of this method is high consumption of solder, more work, and the risk of causing a short circuit between two parallel conducting paths. If you choose this method, be very careful.

3. Lay the wire little by little, until you have created the desired connection.
4. In the conducting path drawing, mark the connection you have just created. That way, when you're finished making the conducting paths, you can easily check whether you made all the necessary connections.

Here's what the completely wired foil side of a circuit board looks like:

*Finished conducting paths*

If you only use a few soldering points to hold the conducting paths, then the foil side of the circuit could also look like the following figure:

*Only a few soldering points were set*

As a general rule, the board is now completely built. However, frequently more work must be done before the entire circuit is complete.

## Additional Work On The Circuit

Often the boards are connected using cable with connectors or similar components. These were not included in the drawing we've talked about so far. To have the circuit function, however, we'll also have to perform these tasks. Remember also, the required components for this additional work are not included in the parts list. Before you buy parts for a circuit, make certain to read through the entire chapter. That way you'll find out which components you need besides those mentioned in the parts list.

For this circuit, you'll need two 25-pin Sub-D connectors, one female and one male (each without a case), approximately 65 centimeters of flexible plastic-coated copper wire and 16 edge connectors. You'll need the connectors to connect the circuit to the port of the computer on one end and to a connected device (e.g., a modem) on the other end.

Now perform the following tasks:

1. Make 16 cable pieces, approximately 1 1/2 inches in length, and strip the insulate approximately 1/2 inch on both ends. Now solder a male edge connector to each end of the cable.

2. Next, take the female Sub-D connector, and place it so that you can see the rear side. As a rule, there will be small numbers there, which number the terminals. You can also find out the terminal assignment in Part I.

3. Now solder one of the prepared cables to terminals 2, 3, 4, 5, 6, 7, 8 and 20. (Make certain to use the free end of the cable.)

4. Follow the same procedure with the male Sub-D connector, taking the same terminals into consideration. Remember, with this connector the numbering goes from right to left.

5. After preparing the connectors, plug the male edge connector into the female edge connector on the board. Use the following allocation.

| Female Connector | | | Male Connector | |
|---|---|---|---|---|
| Wire 7 | Connector 1 | | Wire 7 | Connector 9 |
| Wire 2 | Connector 2 | | Wire 2 | Connector 10 |
| Wire 3 | Connector 3 | | Wire 3 | Connector 11 |
| Wire 4 | Connector 4 | | Wire 4 | Connector 12 |
| Wire 5 | Connector 5 | | Wire 5 | Connector 13 |
| Wire 6 | Connector 6 | | Wire 6 | Connector 14 |
| Wire 8 | Connector 7 | | Wire 8 | Connector 15 |
| Wire 20 | Connector 8 | | Wire 20 | Connector 16 |

6. While we are finished building the circuit, you've got one extra task of attaching the connectors to the board. To be able to plug the circuit properly into the port of the computer, you need to create four small joints. You can screw them onto the connectors and to the board. The connectors usually have the appropriate holes, and you can bore holes into the board.

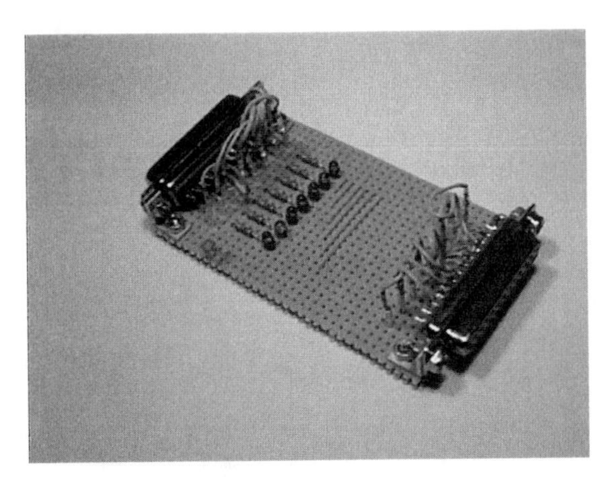

*The finished circuit*

## Using The Circuit With The Companion CD-ROM

As a rule, you'll find programs on the companion CD-ROM for the circuits in this book. These programs are located in subdirectories corresponding to the chapter numbers. The subdirectories also have their own subdirectories. One of the subdirectories contains the installation files, which you can use to install the software on your computer. A different directory contains the source files of the software so you can modify or expand the program to suit your needs.

### No software for this circuit

Since this circuit doesn't allow any influence by a software program, we didn't program any software for this circuit. So, you won't find any software for this circuit on the companion CD-ROM. However, the other circuits do have software programs.

## Using The Circuit

The circuit is ready to use once you've built it. Plug the female connector into your computer's serial port. Next, connect a serial device (e.g., a modem) to the other end of the circuit. Once the device is switched on and data is exchanged over the port, you'll be able to tell by the flickering of the light-emitting diodes.

**Important Tip**

If you don't have a modem handy, you can also connect the circuit to the computer and your mouse. Since data is also exchanged here, the circuit will make the exchange visible.

# Chapter 8:

# Parallel Port Tester

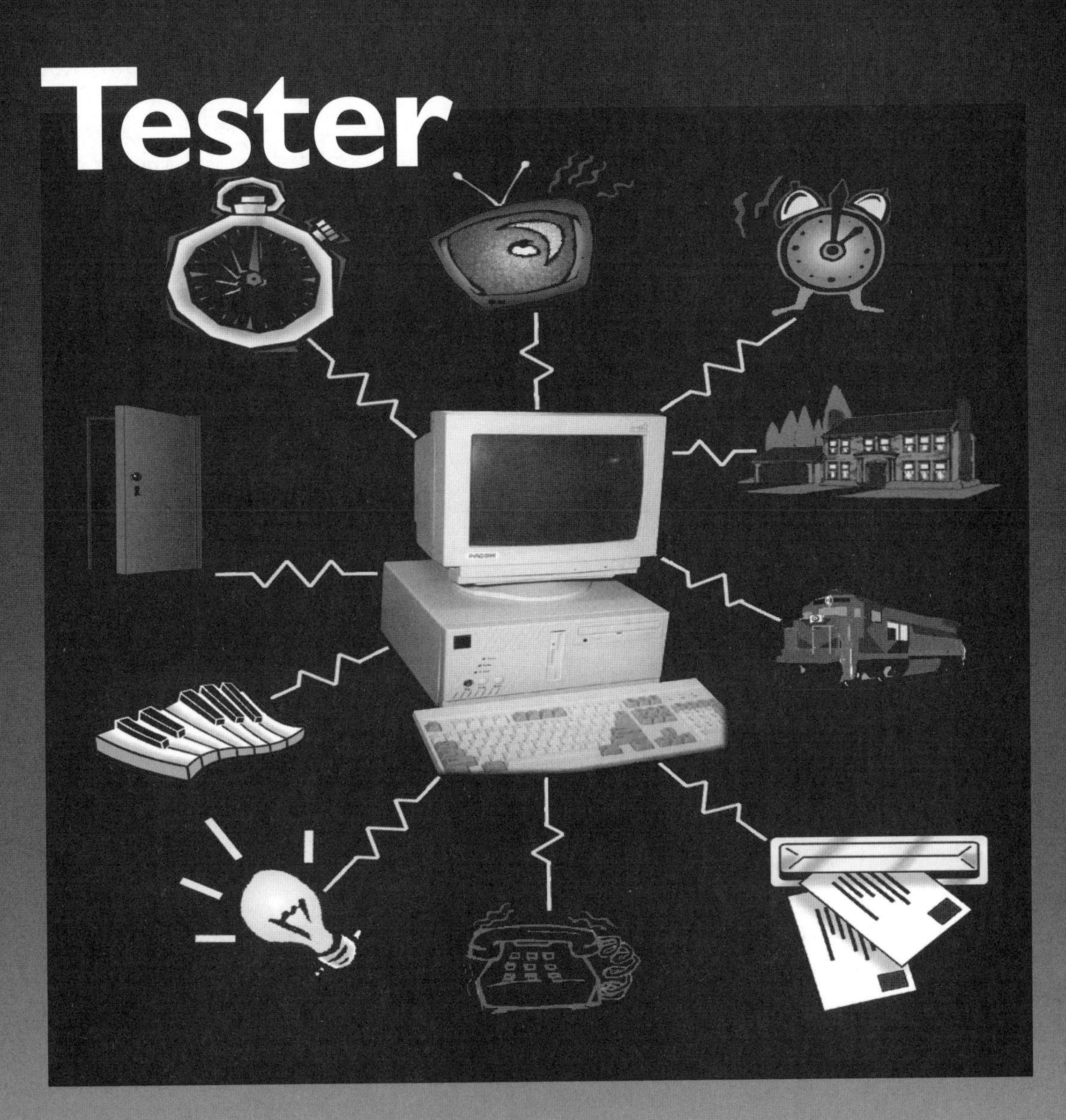

# Chapter 8: Parallel Port Tester

In Part I you learned about the serial port and the parallel port. Chapter 7 talked about a serial port tester and so in this chapter we'll talk about a parallel port tester. You'll also build this circuit on a European circuit board that has been cut in half. In other words, you can use the remaining half from the serial circuit in Chapter 7.

## The Circuit And Its Function

Similar to the serial port tester, this circuit indicates when data and signals are being exchanged over the parallel port of your computer. Naturally, the parallel port operates differently than the serial port; however, the basic principle of how data is exchanged is comparable. The greatest difference is that eight bits are transmitted at the same time in parallel data transfer (data transfer occurs 1 bit at a time with the serial port). The fact that signals change their states in a data transmission, even with the parallel port, is obvious.

The circuit has two main tasks. The first is that the circuit's light-emitting diodes will alert you that data is being exchanged. The circuit should also give you the opportunity to test whether the parallel port will work with the software for this circuit.

The circuit is built to have one side plugged into the parallel connector of your computer and have the other side of the circuit connected to a parallel cable, operating a printer, for example, from the cable without impairing the function.

So, the circuit is inserted between a parallel end device and the parallel port without limiting its function. Since this is only possible when all the wires at the input of the circuit are also available at the output, you will need to do some soldering—another great opportunity for some practice.

Often, the only way to tell whether data is being exchanged is by the flickering of the light-emitting diodes. You can only detect precise line states if you use the software that goes with the circuit. This software makes it possible for you to influence the parallel port in such a way that specific data is set or not set.

## Building The Circuit

As we mentioned, you'll need half of a European ("pre-perfed") circuit board to build the circuit. If you can no longer use the other half of the board you used for the serial port tester (it may have been broken or been used for some other purpose) cut another board in half. Follow the instructions in Chapter 7.

### The location of the components on the board

After preparing the board, turn your attention to the component side of the board. The parallel port circuit contains eleven light-emitting diodes and an equal number of resistors. You'll have to solder these onto the board.

Follow the same procedure you used with the serial port test. For each component, do the following:

1. Find a component from the drawing and use the parts list to determine what kind of component it is. Then get the corresponding component ready.
2. Find the position of the component on the board using the following drawing.
3. Insert the component through the appropriate holes of the board.
4. Turn the board over and solder the component.
5. Use cutting pliers to cut off the excess wire leads.

Use the following drawing when you build the parallel port tester. Use the numbers on the borders of the board to get your bearings.

# 8. Parallel Port Tester

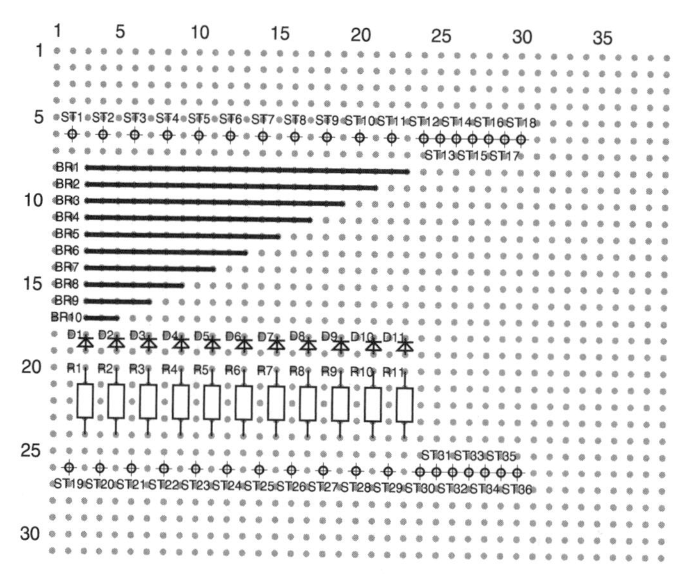


*The location of the components on the component side of the board*

## Parts list

The following parts list shows the individual components you'll be using in this circuit.

| Parts List | | |
|---|---|---|
| Label | Name | Component type |
| BR1 - BR10 | Conducting path bridges | Copper wire |
| R1 - R11 | 1 kΩ | Resistor |
| D1 - D8 | 3-mm light-emitting diode (red) | Light-emitting diode |
| D9 - D11 | 3-mm light-emitting diode (yellow) | Light-emitting diode |
| ST1 - ST36 | Edge connectors | Pin connection |

As was the case with the serial port tester, you'll primarily find resistors and light-emitting diodes here. Only the number differs.

## Building the circuit

Building the circuit is very easy. You will need to pay attention to the polarity of the light-emitting diodes. Do you still remember how the positive pole (anode) is indicated? The longer contact wire is the positive.

Now solder the light-emitting diodes, the resistors and the edge connectors to the board. Once you have completed these tasks, turn the board over and focus your attention on the foil side.

# The Foil Side Of The Board

Between the individual solder joints, the necessary connections to the individual components are still missing.

The following drawing shows the necessary conducting path connections. Remember, the numbers along the borders of the board on the foil side go from top to bottom, but from right to left.

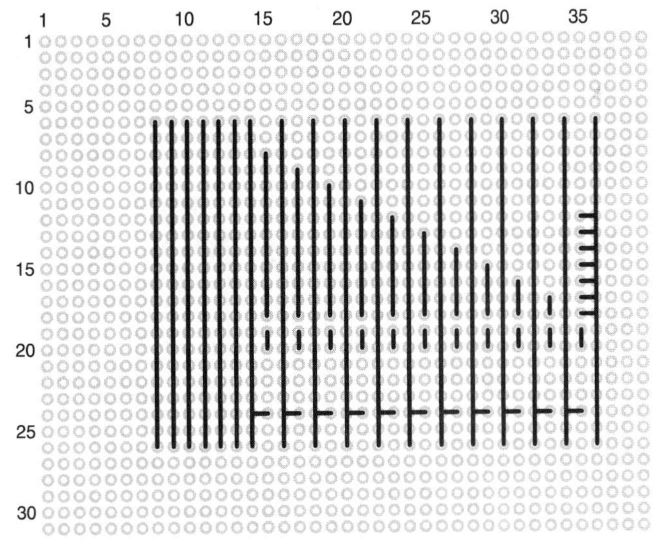


*The conducting paths of the parallel port tester*

After laying the conducting paths, you still have to attach the connectors to the board.

## Additional Work On The Circuit

As was the case for the serial port tester, you will also need two 25-pin Sub-D connectors for the parallel version, one female connector and one male connector (both without case), approximately 4 feet of plastic-coated copper wire and 36 edge connectors.

The connectors are practically an extension of the port, with the circuit between the two.

Attach the connectors as described in Chapter 7. There are only two differences from the circuit described in Chapter 7. The first is that this circuit has several wires that you have to connect. The second difference concerns the parallel tester. The connectors (male and female) are reversed relative to the serial port tester.

**Important Tip**

If you add the extra wires to the connectors of the serial port tester, you can also use them for the parallel circuit. That saves you the trouble of buying two more connectors. However, make sure the wires on the connectors are long enough. Otherwise, follow the arrangement of wires shown in the following table for the circuit.

| Female Connector | | | Male Connector | |
|---|---|---|---|---|
| Wire 18 | Edge connector 1 | | Wire 18 | Edge connector 19 |
| Wire 2 | Edge connector 2 | | Wire 2 | Edge connector 20 |
| Wire 3 | Edge connector 3 | | Wire 3 | Edge connector 21 |
| Wire 4 | Edge connector 4 | | Wire 4 | Edge connector 22 |
| Wire 5 | Edge connector 5 | | Wire 5 | Edge connector 23 |
| Wire 6 | Edge connector 6 | | Wire 6 | Edge connector 24 |
| Wire 7 | Edge connector 7 | | Wire 7 | Edge connector 25 |
| Wire 8 | Edge connector 8 | | Wire 8 | Edge connector 26 |
| Wire 9 | Edge connector 9 | | Wire 9 | Edge connector 27 |
| Wire 1 | Edge connector 10 | | Wire 1 | Edge connector 28 |
| Wire 10 | Edge connector 11 | | Wire 10 | Edge connector 29 |
| Wire 11 | Edge connector 12 | | Wire 11 | Edge connector 30 |
| Wire 12 | Edge connector 13 | | Wire 12 | Edge connector 31 |
| Wire 13 | Edge connector 14 | | Wire 13 | Edge connector 32 |
| Wire 14 | Edge connector 15 | | Wire 14 | Edge connector 33 |
| Wire 15 | Edge connector 16 | | Wire 15 | Edge connector 34 |
| Wire 16 | Edge connector 17 | | Wire 16 | Edge connector 35 |
| Wire 17 | Edge connector 18 | | Wire 17 | Edge connector 36 |

If you made four little joints for the serial port tester, you can also use them here to attach the connectors to the board. Bore the appropriate holes in the board and screw in the connectors, so that you can plug the circuit into the parallel port of your computer.

After you have completed all the connections, the circuit should look like the one shown in the next figure:

*The parallel tester*

## Using The Circuit With The Companion CD-ROM

You'll find the software for the circuit in the *chap_08* subdirectory on the companion CD-ROM. In this directory you'll find a subdirectory called *Install*, which contains an installable version of the software. Start the Setup.exe file and follow the program's instructions. If you intend to change the program, you will find the source code files in the subdirectory of Chap_08. After installation, you can start the program from the Start menu. In the program window that appears, you can enable or disable the check boxes for the individual wires of the parallel port.

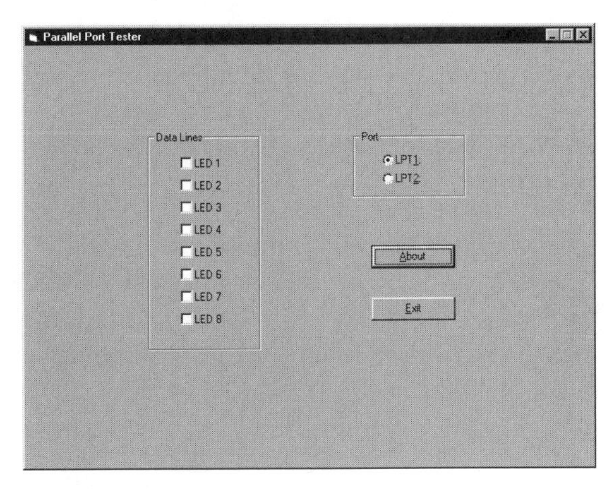


*The software for the parallel port tester*

Depending on which check boxes you enable or disable, you will recognize the effects on the circuit which you have connected to the computer port. In addition, the software lets you choose the port, in case you have two of them installed on your computer. While a computer usually has only one parallel port, you do have the option of selecting the second parallel port, LPT2.

## Using The Circuit

You have already done all the work necessary for operating the circuit. Plug the circuit into your computer's parallel port and start the software program.

**Important Tip**

Make sure the edge connectors don't touch each other. Otherwise, signals could be distorted and the port won't function. This can easily happen since some of the edge connectors are very close to one another. If two connectors do touch each other, use a piece of paper to separate them. Also, make sure no edge connectors touch each other when you are plugging in the circuit or the printer cable.

Change the default port from LPT1 to LPT2 if you use this port, and then click the check boxes for the data lines to enable or disable them. Keep in mind that the parallel port can behave differently, depending on whether or not an end device is connected to the port, for example a printer. Experiment with various settings and find out what changes on the circuit.

Using the software, you can enable or disable the eight red light-emitting diodes (LEDs) separately. These are the eight data lines of the parallel port. The yellow light-emitting diodes are the three signal lines. You cannot influence these with the software, although there will be changes when a connected printer is switched offline or there is no more paper in the printer.

You will also be interested in watching how the signals on the tester change when you output something on the printer using a software application. Remember to exit the software for the circuit before running the application.

With the help of this circuit, you can easily tell when data is being sent over the parallel port. What is more, you can also use the circuit software to determine whether your port functions properly with the software.

# Chapter 9:

# Programmable Surface Light

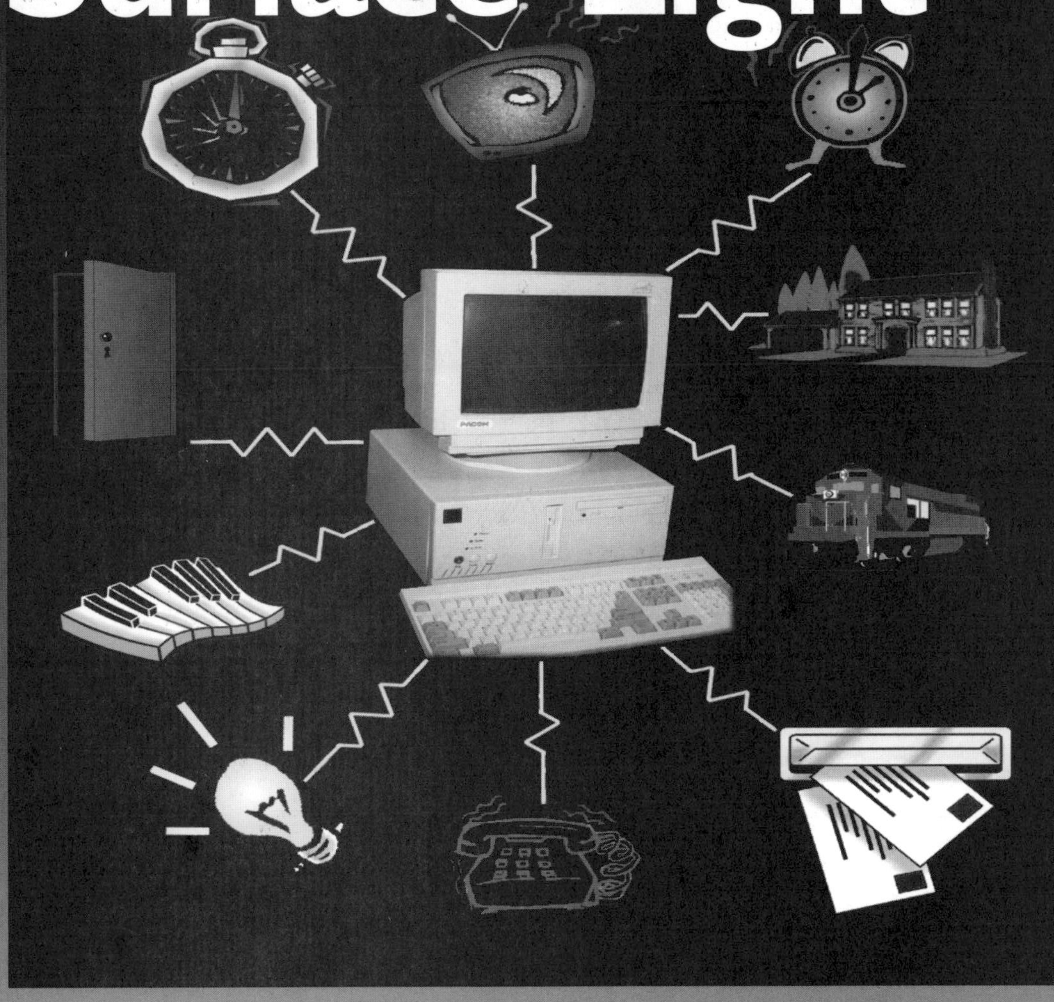

# Chapter 9: Programmable Surface Light

The circuits in Chapters 7 and 8 showed that your PC exchanges a wealth of data over the serial and parallel ports. Using the light-emitting diodes, you can recognize the various line states. One of the two circuits operated without any additional software.

This chapter's circuit is a surface light controlled by the circuit's software. The surface light consists of a series of light-emitting diodes that light up in different ways and sequences, thus delivering varying optical effects.

However, along with the function of this circuit, this chapter also includes another special feature: we discuss and build a motherboard that will be used with the surface light. Several circuits in Part II also work with this motherboard. The motherboard is the basis for the circuits. The add-on boards that you plug into this motherboard provide the different functions.

There are several reasons for using the motherboard. You can save yourself some work since you only have to build this board once, as opposed to building it anew for each circuit. Another reason is that you'll be conserving materials, which, in turn, saves you money.

There are still more reasons for using this motherboard. It's purpose is to relieve the load from the port of the computer. This is done by supplying the motherboard with additional voltage. In this way, you can use components that require more current than the computer port can deliver.

## The Circuit And Its Function

As we mentioned, the actual function of the circuit will be achieved through an add-on board. This board contains a number of light-emitting diodes placed next to one another. A program on the companion CD-ROM controls these light-emitting diodes in various ways.

You could use the surface light in your work room or in conjunction with a train set.

We intentionally kept the circuit simple, so that this chapter will allow you to further delve into the material and become comfortable with it.

## Building The Circuit

There are two phases in building this circuit. First, you will build the motherboard. Then you'll create the add-on board.

## The Motherboard

You'll build the motherboard on a European ("pre-perfed") standard circuit board. One side of it contains a connection option for a normal printer cable so it can be connected to the parallel port of the computer. The other side the circuit contains an edgeboard connection that receives the add-on circuits.

In addition, there will also be a connection for the external power supply, which, in conjunction with the electronics on the circuit, supplies additional current for the plug-on boards. The motherboard contains electronic components, which forward the signals from the port to the add-on board.

The following drawing shows the location of the individual components on the board. Solder the components carefully to the board, in accordance with the positions shown on the drawing.

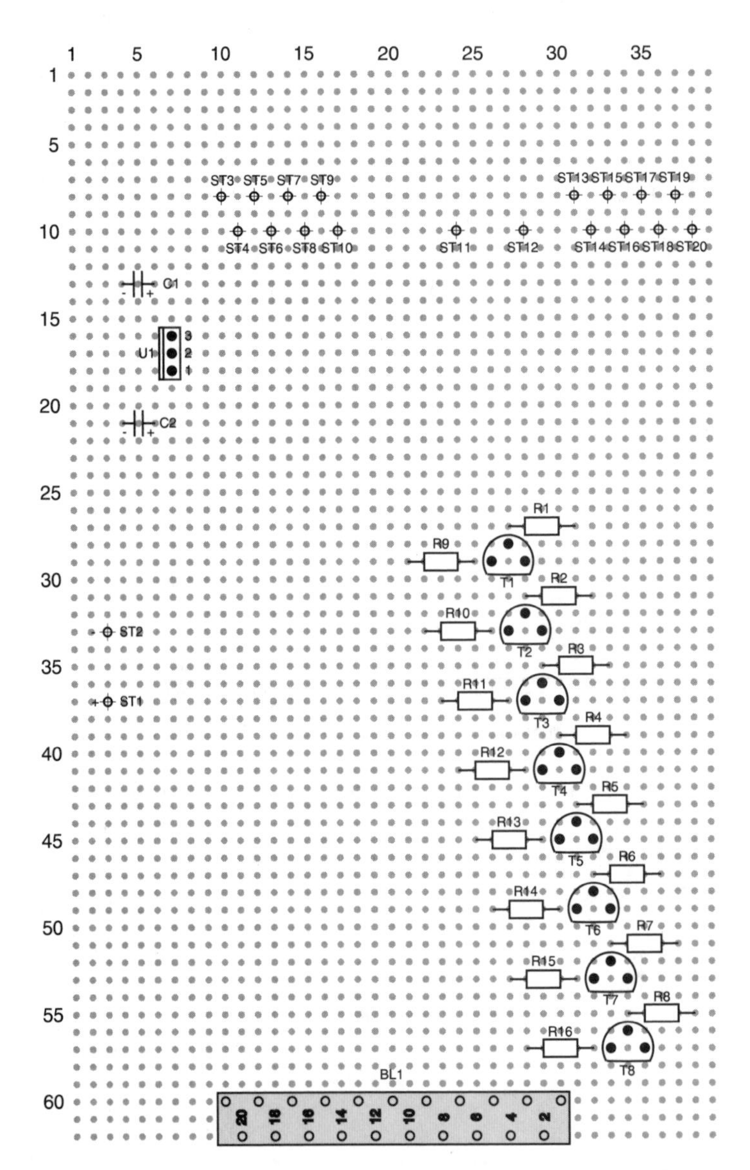


*The location of the components on the motherboard*

When attaching the socket strip, make sure you plug it in correctly, and that it rests completely on the board. You won't have any trouble with the other components.

## Parts list

We created the parts list in the following table to help you coordinate the components in the drawing. Before soldering, place the components in the positions shown in the drawing.

| Parts List | | |
|---|---|---|
| Label | Name | Component Type |
| R1 - R8 | 1 kΩ | Resistor |
| R9 - R16 | 180Ω | Resistor |
| C1, C2 | 100 nF | Capacitor |
| U1 | 7805 | Voltage regulator |
| T1 - T8 | BC 548 | Transistor |
| BL1 | Socket strip 21pin | Pin connection |
| ST1 - ST20 | Edge connectors | Pin connection |

Creating the component connections on the foil side of the board represents more work than the soldering of the components. Because we're working on a full-size board with many components, we'll have to lay a great deal of conducting path wire.

However, this can be a positive experience. After performing this work, you'll have so much practical experience in soldering conducting paths that this type of work will no longer be difficult when you build the other circuits in this book.

## The foil side of the basic circuit

Create the conducting path connections as shown in the following drawing.

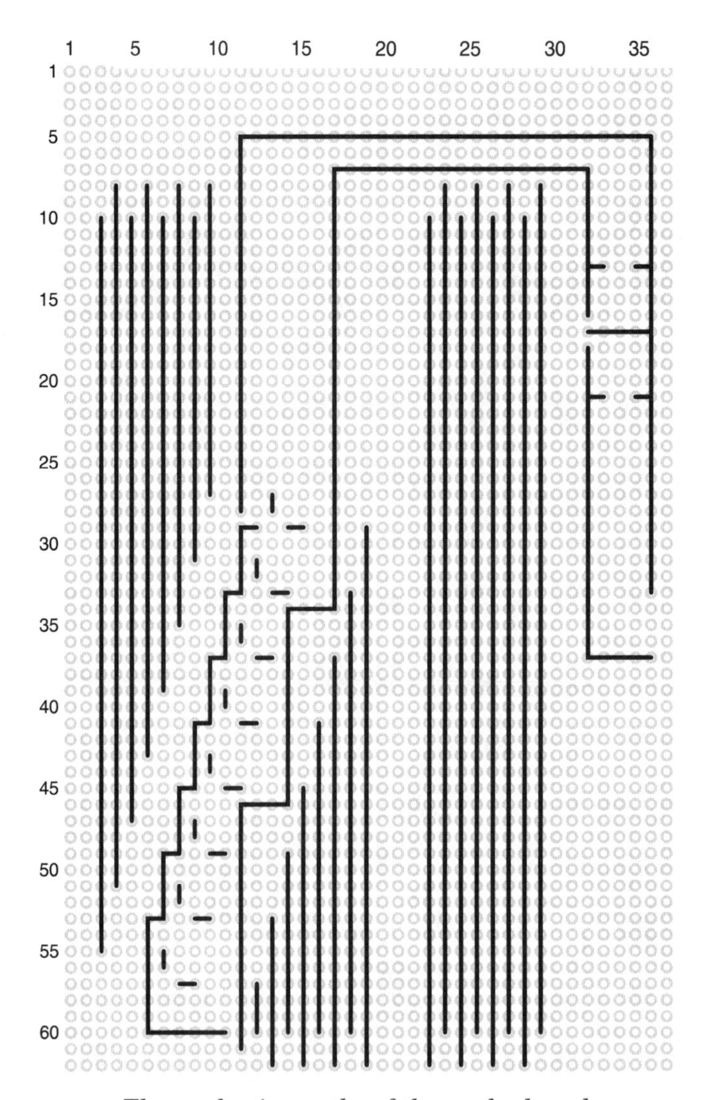


*The conducting paths of the motherboard*

Before you can use the basic circuit, there are two additional jobs to perform. First, create the connection to an external power supply.

The power supply should satisfy the requirements in Chapter 4 ("The Materials You'll Be Using"). As a rule, in addition to the connection for the normal socket, the power supply also has a cable with a plug or assortment of plugs attached to it.

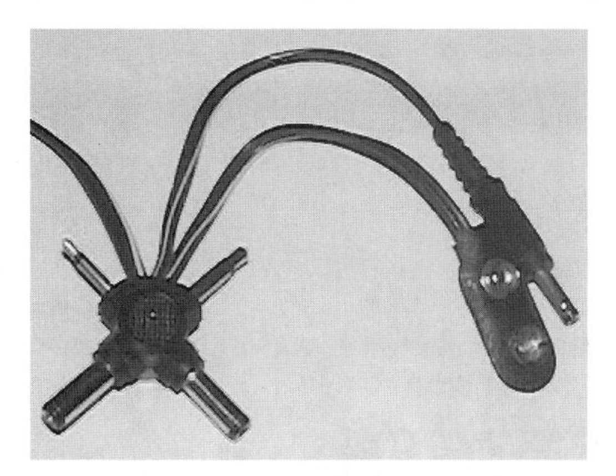

*The various adapter plugs for direct current voltage on the power supply*

On this end of the cable the desired direct current voltage is provided. The direct current voltage must now be connected to the two (female) edge connectors, ST1 and ST2, on the board. Use male edge connectors, so that the power supply can be easily removed from the circuit and used elsewhere if necessary.

One way to achieve this is to disconnect the plugs shown in the illustration above and directly solder the male edge connectors to the two conductors of the direct current cable. Since this means that the power supply won't be preserved in its original state, this option won't appeal to everyone.

As an alternative, you can purchase the plug's counterpart in the electronics shop and solder a short piece of cable to it, on whose end you can then attach the male edge connectors. Then you can plug the *adapter cable* to the plug to connect it to the board.

With both methods, make sure that you can localize the positive or the negative plug at any time. It's best if you mark the positive plug so that you can easily recognize it.

At all costs, avoid the possibility of the two edge connectors touching one another, thus causing a short circuit. While the power supply will usually survive this without damages, it is something that ought not happen.

To avoid a short circuit, it makes good sense to insulate both male edge connectors. For example, you can do this with adhesive tape. Heat shrink tubing offers a much better solution. You slide it over the connectors, and when heated it contracts until it is seated firmly on the connectors. You plug it over the connectors, and when heated it contracts until it is seated firmly on the connectors.

It's up to you which method you choose. Once the power supply has been properly prepared, you can create the connection to the external power supply by plugging the male edge connectors onto the female connectors. Remember to plug the positive plug onto ST1 and the negative plug onto ST2.

**Important Tip**

Don't plug the power supply into the socket until you finish building a circuit and have made all the necessary connections to the computer and the other components. It's too easy to do damage otherwise.

Beside the connection to the power supply, the circuit must also be able to connect to the parallel port of the computer. Since in most cases a printer is connected to this port, quite often a printer cable will be present. Use this cable for this circuit and the ones that follow. In this way you can build and use the circuit as far away from the computer as the cable is long, and connections to the circuits can be quickly made or broken.

A printer cable has a plug on one end that fits into the parallel port socket of the computer. On the other end of the cable is a Centronics plug that fits into the appropriate socket of the printer.

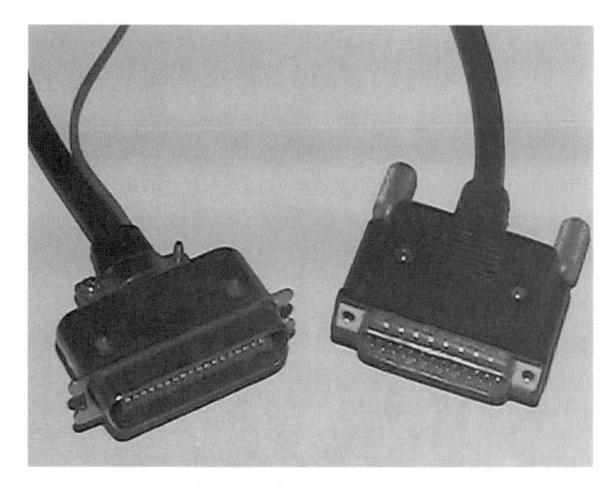

*The connectors of a printer cable*

The idea is to connect the circuit to this Centronics plug in place of the printer. To do this you will need the right jack, which you can purchase at your dealer. This jack must fit into the 36-pin Centronics plug of the cable on one side, while being able to accept a flat ribbon cable on the other.

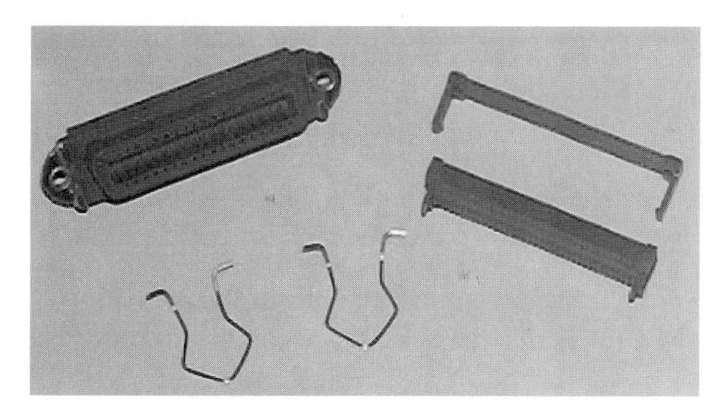

*The individual parts of the Centronics jack*

The figure shows the required jack. To allow connection of a flat ribbon cable to this jack, you will need a model. This has two rows of pins that are forced into the individual conductors.

Use the following steps to connect the flat ribbon cable to the jack:

1. Place a 16 to 20 inch long 36-pin flat ribbon cable next to the Centronics jack. Place it on the pin rows of the jack in such a way that each conductor is over a pin. One end of the cable should jut out about 1 cm beyond the jack. Then press the individual leads into the pin columns from one side to the other. The pins must cut through the insulating material of the cable and be visible on the top side of the flat ribbon cable. The copper wires create a connection to each individual pin.

**Important Tip**

Locate contact 1 on the jack. The numbers are often difficult to make out, but they are there. If you are using a flat ribbon cable whose first conductor is marked in color, then plug it into the jack so that the marked wire is connected with contact 1. If none of the conductors are marked, change this and make this wire recognizable, for example by marking it with a felt marker. You can also locate contact 1 by referring to the drawings in Chapter 5.

2. Once the flat ribbon cable is plugged in, use one of the plastic clips attached to the jack to hold it in place. You don't want it to slip out of the pin columns. This sturdy clip must press firmly on the flat ribbon cable and connect firmly with the jack through a locking device.

3. Next, use a pair of scissors or cutting pliers to snip off the excess length of the flat ribbon cable directly at the jack, so that it's not in your way.

4. Bend the flat ribbon cable protruding from the jack directly behind the clip carefully by 180° to the top beyond the clip to the back. The cable must lie flat on the clip.

5. Then place the second, somewhat narrower clip attached to the jack over the bent flat ribbon cable on the first clip. As soon as the second clip snaps firmly into place, all you have to do is stick the two metal clips into their openings, and you are finished attaching the jack.

*The Centronics jack connected to the flat ribbon cable*

Now you can connect the flat ribbon cable to the printer cable. To connect the other end to the circuit, you'll have to solder male edge connectors to some of the wires.

To do this, spread the individual wires of the flat ribbon cable at a length of about 5 cm from each other, so that individual wires are available. Note the marked wire number 1. Note wires 1, 2, 3, 5, 7, 9, 11, 13, 15, 17, 19, 21, 23, 25, 27, 28 and 36, starting your count with wire 1 and increasing the count by 1 for each wire. Therefore, don't pay any more attention to the contact numbering on the Centronics jack. Strip the ends of the wires about half an inch, and solder a male edge connector to each one.

Now you can make the necessary connection to the circuit using the edge connectors. Consult the following table to find out which wire goes with which plug.

| | | | | | | | |
|---|---|---|---|---|---|---|---|
| Wire 1 | ST3 | | Wire 11 | ST16 | | Wire 23 | ST8 |
| Wire 2 | ST12 | | Wire 13 | ST15 | | Wire 25 | ST7 |
| Wire 3 | ST20 | | Wire 15 | ST14 | | Wire 27 | ST4 |
| Wire 5 | ST19 | | Wire 17 | ST13 | | Wire 28 | ST5 |
| Wire 7 | ST18 | | Wire 19 | ST9 | | Wire 36 | ST6 |
| Wire 9 | ST17 | | Wire 21 | ST10 | | | |

Once the wires are attached, the basic circuit is ready to be used.

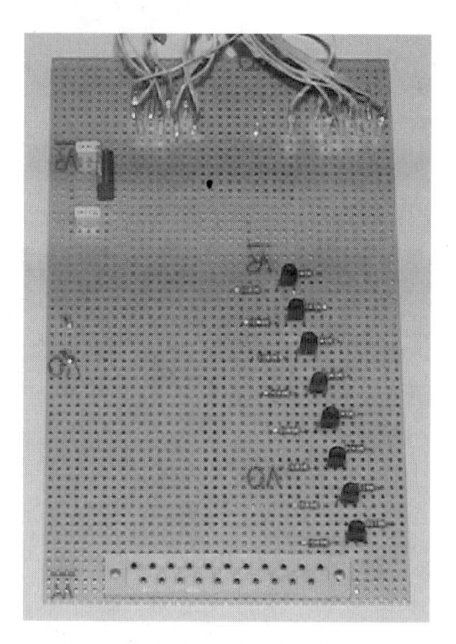

*The finished basic circuit*

## The Add-on Board

After you finish building the motherboard, you are ready to turn your attention to the surface light. As we explained earlier, you build the surface light on a separate board, which is then plugged into the socket strip of the motherboard.

Along with the edgeboard connection, there are only eight light-emitting diodes to be soldered. See the following drawing to determine their positions on the board.

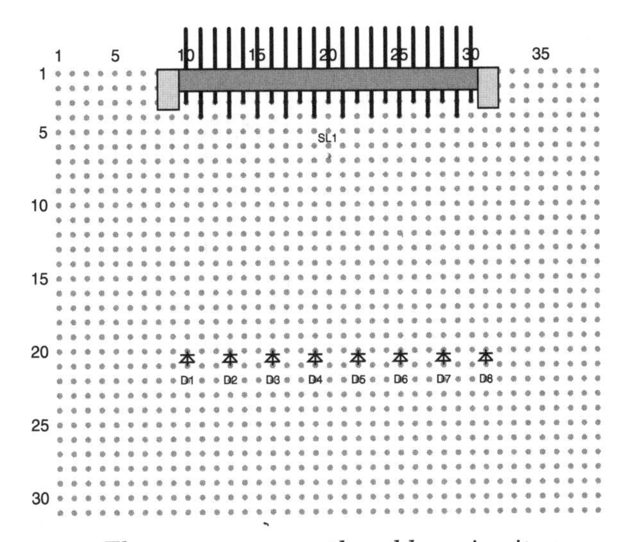


*The components on the add-on circuit*

When you're soldering, pay attention to the polarity of the light-emitting diodes and make sure the edgeboard connection is seated firmly on the board.

## Parts list

The following table shows the parts list.

| Parts List | | |
|---|---|---|
| Label | Name | Component type |
| D1 - D8 | 5-mm light-emitting diode (red) | Light emitting diode |
| SL1 | edgeboard connection 21pin | Pin connection |

# The Foil Side Of The Board

Make the necessary component connections on the foil side of the board. Consult the drawing below.

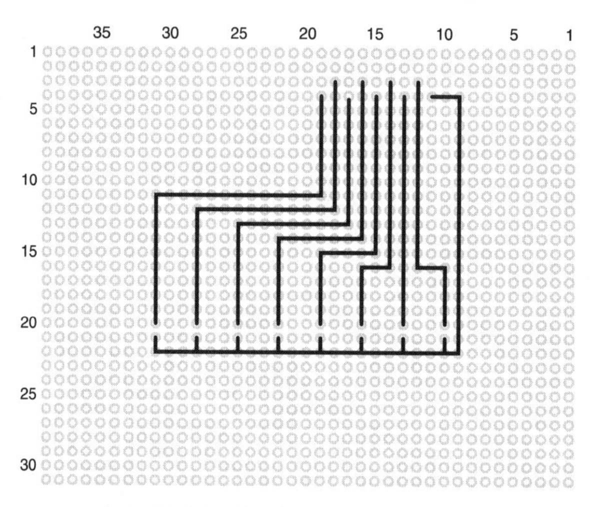


*The Foil Side of the Surface Light*

You can see that building the basic circuit was much more time consuming than making the add-on circuit. You will notice the same pattern with some of the other circuits in this part of the book.

*The add-on board*

## Using The Circuit With The Companion CD-ROM

You are finished building the hardware of the surface light. However, without software this hardware won't do you any good at all. It's the software that controls the light-emitting diodes through the basic circuit, thus achieving an optical effect.

The companion CD-ROM has the appropriate software. You will find all the files necessary for installing the software on your computer in the directory entitled *chap_09*.

If you plan on customizing the software to suit your own requirements, you can do this by loading the source files from the *chap_09/source* directory in Visual Basic and making the necessary changes.

After installation you can start up the software. The following window appears.

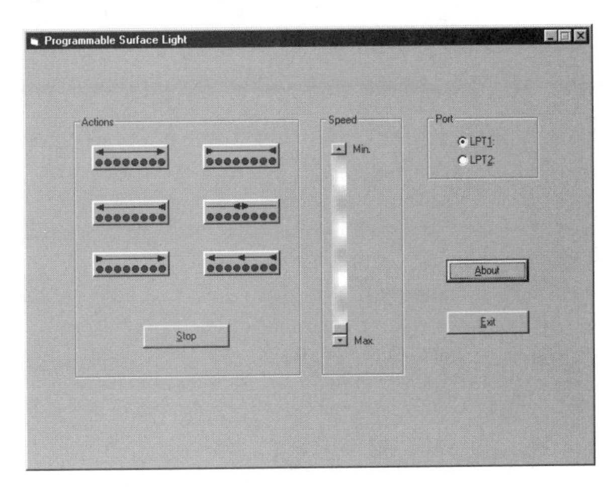


*The software for the surface light*

The *Port* area gives you the option of choosing the port from which you will run the circuit.

In addition, the *Actions* area has various buttons that you can select. Each button influences the surface light. Once you have selected a button, the surface light will respond in a continuous loop. You won't be able to choose any other action button until you click the *Stop* button to stop the current action.

Along with toggling the light-emitting diodes on and off, you can also change the *Speed* at which the light-emitting diodes are controlled. To do this, move the *Speed* slider in the desired direction.

Experiment with the different options.

Since we kept the programming very simple, you will easily understand how to control the individual light-emitting diodes. If you get the opportunity, try to link further actions. A number of optical effects are conceivable.

## Using The Circuit

Operating the circuit won't be difficult if you have followed the previous work steps. To operate the circuit, follow these instructions :

1. Plug the surface light add-on board with the edgeboard connection into the socket strip of the motherboard.
2. Using a printer cable, connect the basic circuit to the parallel port LPT1 of your computer.
3. Plug the external power supply into the motherboard, check whether the power supply is set to 9 volts and the polarity is correct, and plug the power supply into the socket.

If you have installed the software, all you need to do is start it and the circuit is ready to use.

# Chapter 10:

# The Light Bulb Tester

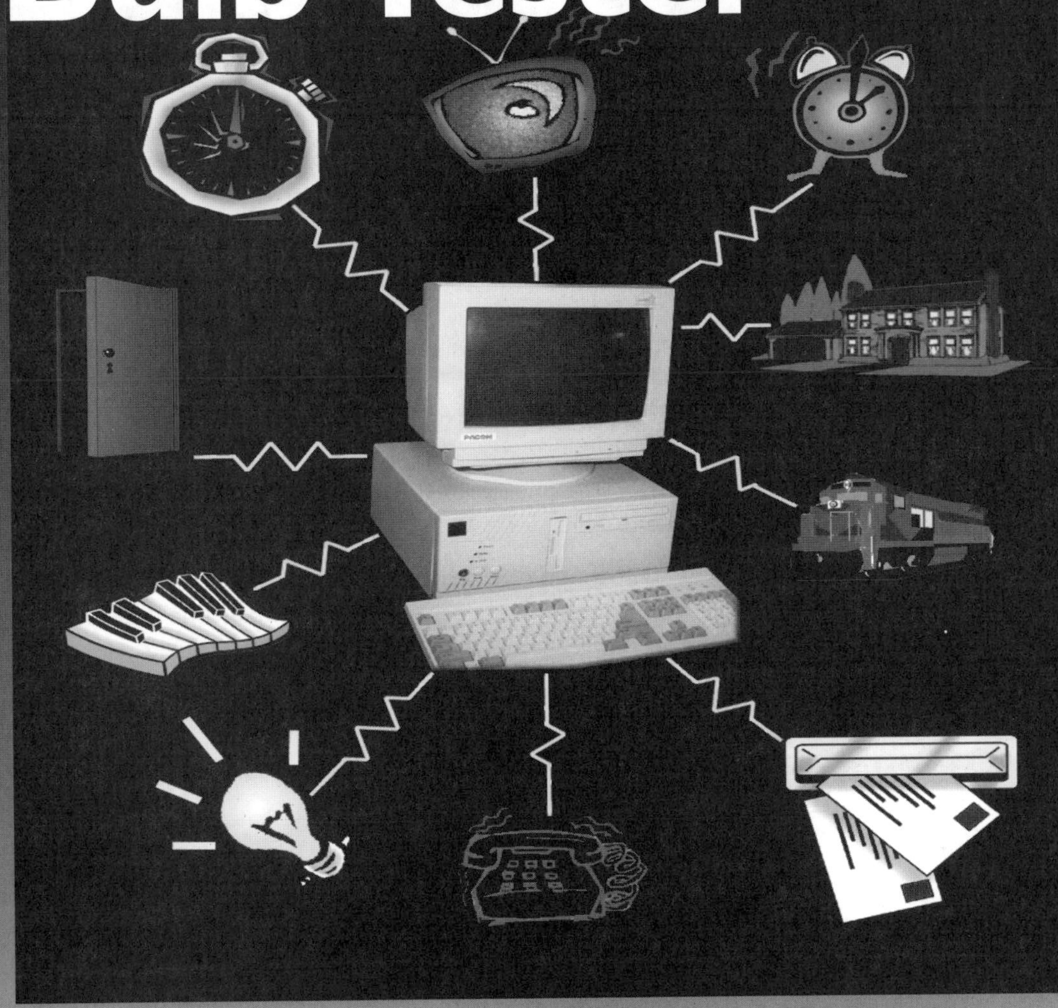

# Chapter 10: The Light Bulb Tester

You've probably put light bulbs away in a drawer or on a shelf in the storage closet. Then when you come back at a later time needing a light bulb, you're not sure whether the bulb still works. The circuit you build in this chapter lets you check the light bulbs you use in your home to see whether they're still good. This circuit and its software on the companion CD-ROM will help you easily answer this question, no matter what type of light bulb it is.

This circuit works with the basic circuit you built in Chapter 9. You will need half a European standard circuit board.

## The Circuit And Its Function

The function of the light bulb tester is quite simple. Two contacts are attached on the circuit board so they can easily form a connection to the two contacts of a light bulb. The circuit determines whether the filament of the light bulb is still intact.

The software analyzes the result and displays it on the monitor, both visually and audibly. In addition, a green light-emitting diode will light up on the circuit so you can see the results of the test there as well.

## Building The Circuit

As we mentioned, this circuit requires the basic circuit from Chapter 9. You'll plug this circuit into the socket strip on the basic circuit.

You will build the circuit on half of a standard European ("pre-perfed") circuit board. If you don't have any more half boards on hand, you will have to cut one in half. You will also need a sturdy wire for the circuit, approximately 3 to 4 inches long. This wire should have a cross-section of approximately 14 gauge.

The components are arranged as shown in the following drawing.

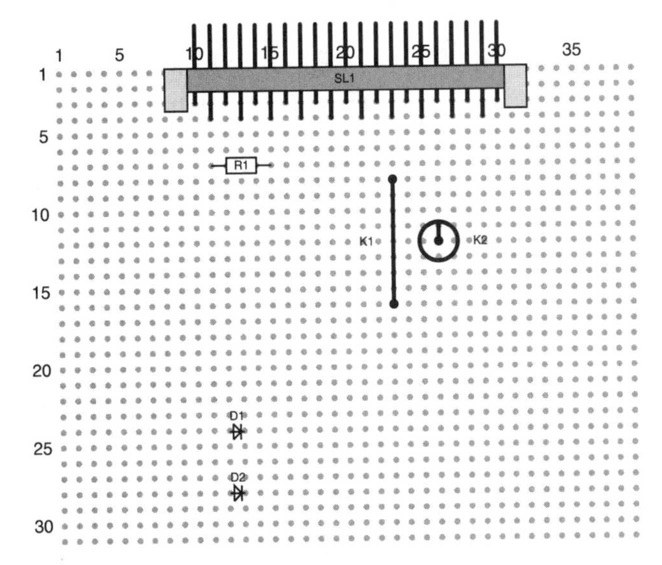


*The component side of the circuit*

## Parts list

As you can see in the following parts list table, you don't need many components for this circuit.

| Parts List | | |
|---|---|---|
| Label | Name | Component type |
| R1 | 10 kΩ | Resistor |
| D1 | Light-emitting diode 5 mm (red) | Light-emitting diode |
| D2 | Light-emitting diode 5 mm (green) | Light-emitting diode |
| K1, K2 | Contact wire | Wire 1.5 $mm^2$ cross-section |
| SL1 | Pin strip 21pin | Edgeboard connection |

### Building the circuit

Before soldering the components to the board, prepare the contact wires.

1. To do this, take the wire with the 14 gauge cross-section and strip it, if you haven't already done so. Then cut off a piece that is 6 centimeters long. You will use this piece for contact K1.

2. Using a pair of pliers, bend this piece of wire approximately 1 inch from one end downward by 90°. At the other end of the wire, form a 'U' shape by bending a piece of equal length downward by 90 degrees.

3. Use the remaining piece of wire for contact K2. Twist this piece in the shape of a circle. Once you have created the circle, bend one end of the wire down, so that 1/8 to 1/4 of the wire is at a 90° angle to the circle.

4. After preparing the materials, locate the appropriate holes in which the contacts are to be inserted. You may need to expand these holes a bit, for example, with a 1/8 inch drill. First insert the bent end of contact K2 into the appropriate hole, and solder it on the rear side. Then insert bow contact K1 into the two other holes and solder it as well.

5. After that you can solder the remaining components. Make certain that the edgeboard connection lies flat on the board. Pay attention to the polarity of the light-emitting diodes.

When the components are soldered, you'll need to turn your attention to the connections on the foil side of the board.

## The Foil Side of The Board

Install the conducting paths on the back side of the board as shown in the following drawing.

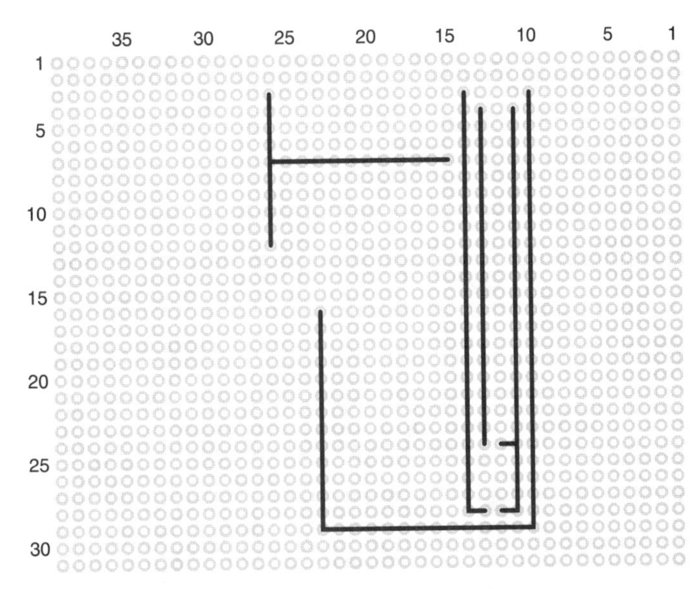


*The foil side of the circuit*

There's nothing special to remember when you install the conducting paths. However, make certain the contacts aren't twisted after you solder the wires.

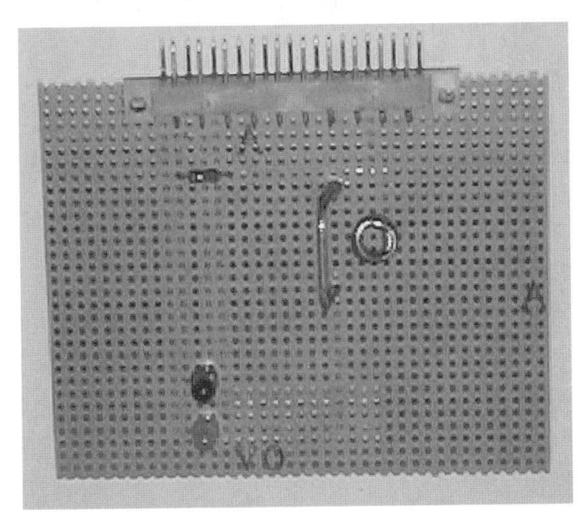

*The finished circuit*

## Using The Circuit With The Companion CD-ROM

To use the circuit, you will need the software from the companion CD-ROM. You will find this software in the *chap10* directory. If you wish to make changes to the software, you will find the original files in the *Source* directory.

After loading the program, you are ready to start it. After test mode starts, you will see the following screen:

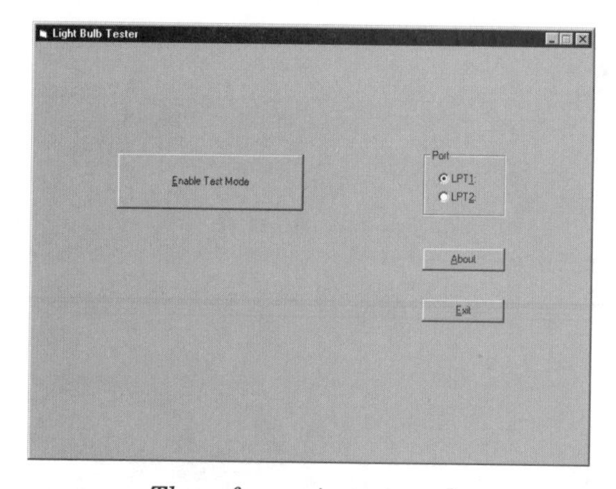


*The software in test mode*

After clicking the *Enable Test* button, the software continuously checks the status of the circuit. As soon as you hold a light bulb that is in working order on the contacts, the light bulb icon will turn into a flashing light bulb. An audio signal is also output. On the circuit, the green light-emitting diode will light up.

If the light bulb to be tested is defective, no signal will be output. On the circuit, the red light-emitting diode will continue giving off light, indicating that the circuit is running in test mode.

## Using The Circuit

Use the circuit as described below.

1. Plug the circuit into the basic circuit from Chapter 9. Now connect it to the parallel port of your computer. Supply the circuit with power from the power supply.

2. Now start the software and click on the *Enable Test Mode* button.

3. Test the function of the circuit by using a screwdriver to create a connection between the two contacts on the board. The software should output an visual and audible signal. After that, the circuit is ready to use.

4. To test a light bulb, hold it against the two contacts of the circuit so that a connection to the two contacts of the light bulb is created.

# Chapter 11:

# The Continuity Tester

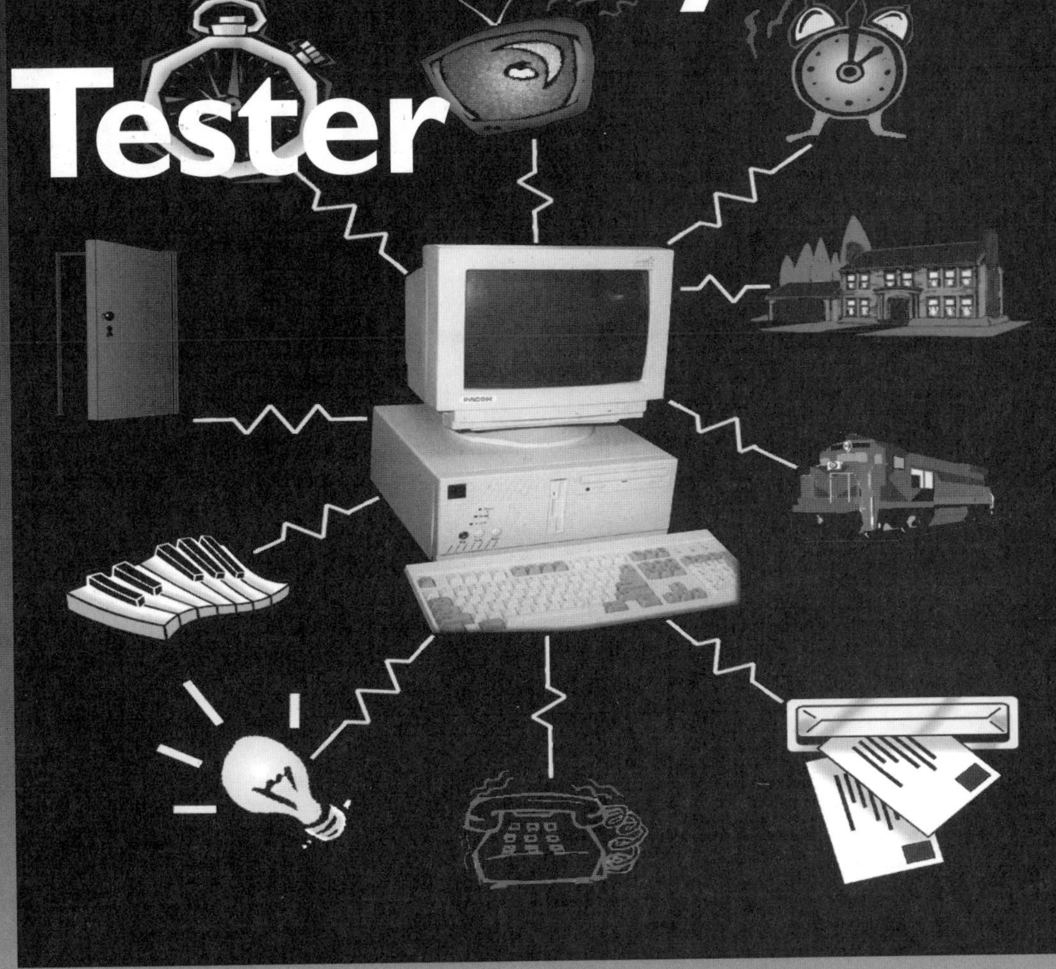

# Chapter 11: The Continuity Tester

The circuit in this chapter is similar to the circuit in Chapter 10. It also works with the basic circuit from Chapter 9.

The title of the chapter should give you a good idea of the function of this circuit. Cable is always being laid for household appliances or hobby uses. Often, users aren't sure which cable end on one side matches a cable end on the other side.

If the cables are marked in color or by pattern, making a connection is not very difficult. However, it's a different story when there is no such identification.

This circuit was designed to help you identify cables. In addition, you can use this circuit to determine whether there is an electrical connection between two points.

## The Circuit And Its Function

The circuit works according to a very simple principle. Two male edge connectors are attached to the board. You can attach a cable to each of these, on whose end you can connect measuring tips or terminals. You can then connect these to the test points to be checked.

Once you connect the circuit to your computer and boot the program, it will check whether there is an electrical connection between the two test points. If there is a connection, the software gives you both a visual and an audible message. Otherwise, the program won't react if no connection occurred.

**Important Tip**

The electrical continuity tester cannot be used within an electronic circuit as a measuring device. For this purpose, you would have to resort to devices made for this purpose, such as a device for measuring voltage. Also, you can only use the continuity tester when the wires to be measured are not live.

## Building The Circuit

The circuit is an add-on board that plugs into the basic circuit from Chapter 9. To build the circuit, you will need half of a standard European ("pre-perfed") board. You also need two flexible cables to furnish the contact to the two test points. The length of the cable is determined by the distance of the two test points from the computer.

You will build the circuit as shown in the following drawing.

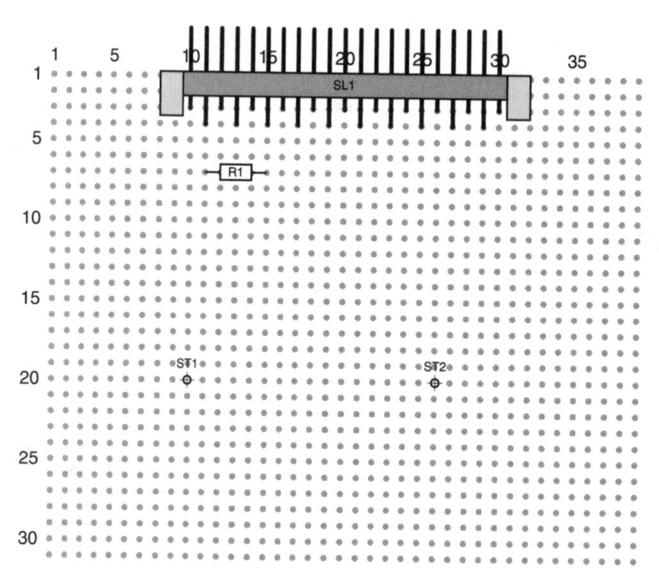


*The design of the circuit*

## Parts list

The following table lists the components you will need.

| Parts List | | |
|---|---|---|
| **Label** | **Name** | **Component type** |
| R1 | 10 kΩ | Resistor |
| ST1, ST2 | Edge connectors | Pin connection |
| SL1 | Edgeboard connection 21 pin | Edgeboard connection |

As you can see, the circuit consists of only a few components so building the circuit is easy. Place the components in line with the above illustration. No special steps are required.

Connect the components as we'll describe in the next section.

## The Foil Side Of The Circuit

Because the circuit consists of just a few components, you'll only have to install a few conducting paths on the rear side of the board. Consult the following figure for the required connections.

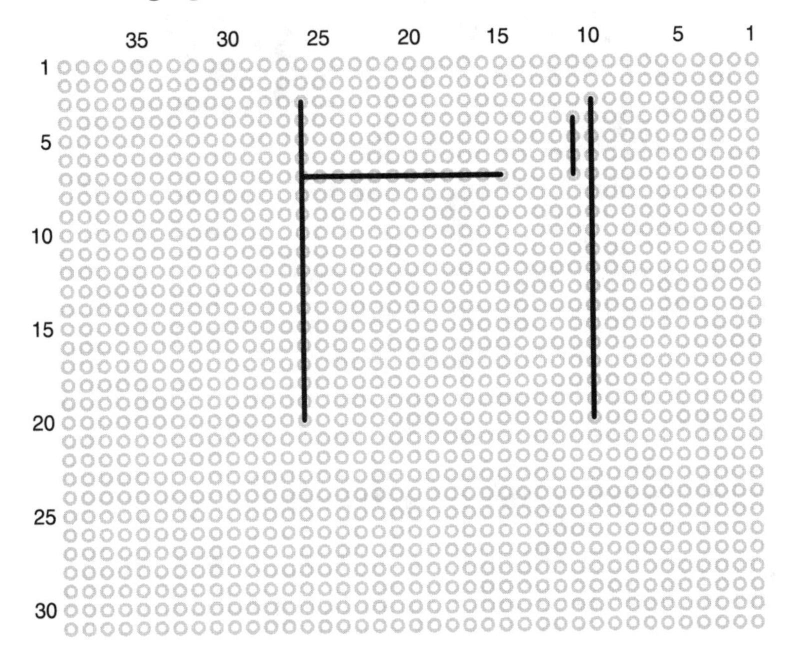


*The foil side of the circuit*

There's nothing special to keep in mind when you install the conducting paths.

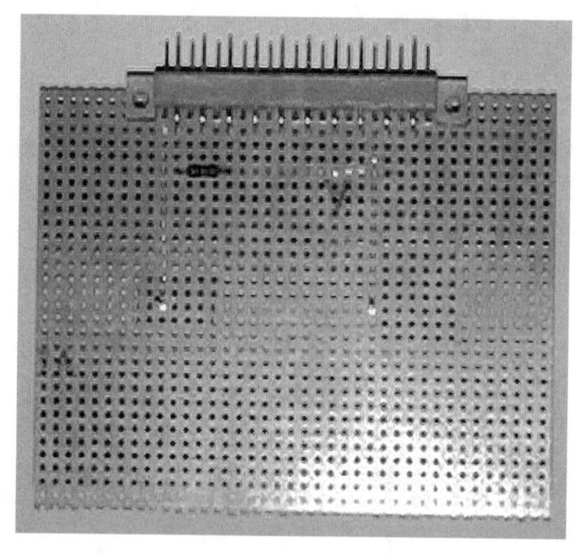

*The finished circuit*

## Additional Work On The Circuit

After building the circuit, you can plug it into the edgeboard connection of the basic circuit. You still need the measuring cables of the circuit.

As we mentioned earlier, you don't need special cable for the measuring cable. Bell wire will suffice, and doesn't cost very much. Since there won't be any great currents flowing through the measuring cable, the cables won't need a large cross section either.

However, there's one item to keep in mind. Since the cables are frequently going to be bent when they are used, the cable should be made of *stranded wire*. Stranded wire is not made up of a single massive copper wire, but instead, is made up of many very thin copper wires. This makes the cable much more flexible and less apt to break.

Once you have purchased the measuring cable, you'll have to do some work with it. However, for this circuit, it's not all that much extra work.

1. Solder a male edge connector on one end of each of the two measuring cables. Only then can you connect the measuring cables properly to the circuit. To do this, remove approximately three-fourths to one inch of insulation from one end of each of the measuring cables Then solder the male edge connectors to these two cable ends. After that, plug the cables into the two female edge connectors of the circuit. Since you don't need to worry about polarity, you can plug the cables anywhere.

2. On the other two ends of the measuring cables, it's a good idea to attach something to create the connection to the test points. One option would be to strip the two ends and place the wires coming out of the cables on the test points, thus achieving the connection. However, this solution is unsightly and also ineffective, since the cable could slip, especially if the two test points are far apart from one another.

   *Alligator clips* are a good solution. You can attach them to the two free ends of the measuring cables. Crimp or solder the clips onto a stripped piece of cable. A combination of the two is the best approach for this situation.

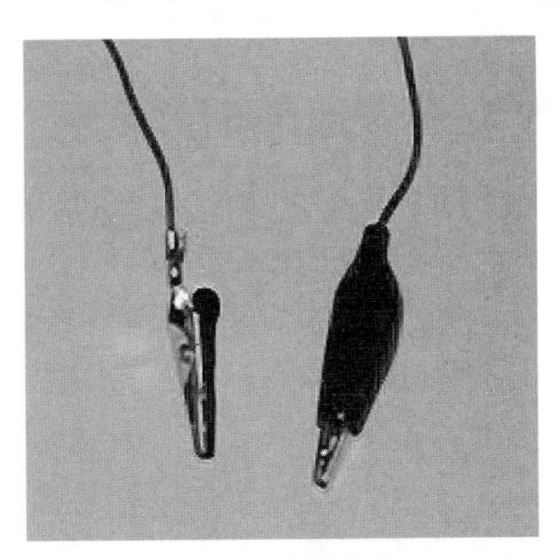

*Measuring wires with alligator clips*

Once the alligator clips are attached, you can clip them onto the test points, just like clothespins.

## Using The Circuit With The Companion CD-ROM

You'll find the program for this circuit on the companion CD-ROM. The two subdirectories with the installable program and the Visual Basic source files are located in the *chap_11* directory.

To begin installation, run the *Setup.exe* file. You are already familiar with the rest of the procedure. As soon as you finish installing the software, you are ready to start the program.

The software doesn't automatically begin testing after startup. You have to click the *Enable Test Mode* button, then it will continuously test the state between the two measuring cables. Once an electrical connection exists between the two cables, the program displays this on the monitor. In addition, it outputs an audio signal, so that you can recognize the results of your measurement without having to look at the monitor.

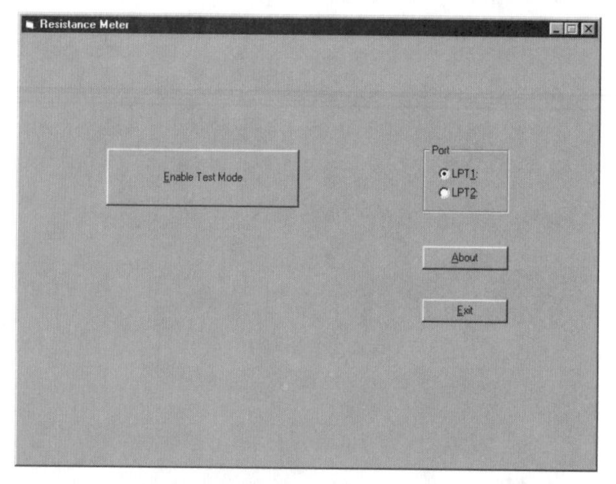


*The program in test mode*

If you remove the cables from the test points, the program reverts to test mode. Click the button a second time (the button now reads *Disable Test Mode)* to switch off the testing function of the software.

You can easily test whether the program is functioning properly with the circuit by holding the two ends of the measuring cables to each other. The program should immediately output the optical or acoustic signal.

If you're operating the circuit at a second parallel port (LPT2), you must communicate this to the program by selecting the appropriate option.

## Using The Circuit

After building the circuit and installing the program, operate the circuit as described below.

1. First, plug the circuit into the basic circuit from Chapter 9. Then connect the circuit to the parallel port of the computer. Furnish the circuit with power through the external power supply.
2. Next, start the program.
3. To test whether the circuit is functioning properly, hold the two cable ends to each other. You will be able to see or hear the results of this test.
4. To test a connection, connect one end of a cable into one of the test points, and hold the other cable to the other test point. If there is an electrical connection between the two test points, the visual and audio signals will appear.

**Important Tip**

Never measure cables that are live, e.g., between the two contacts of a socket. Serious injury could result if you touch these cables. At the every least, you could destroy the circuit.

# Chapter 12: Electronic Dice

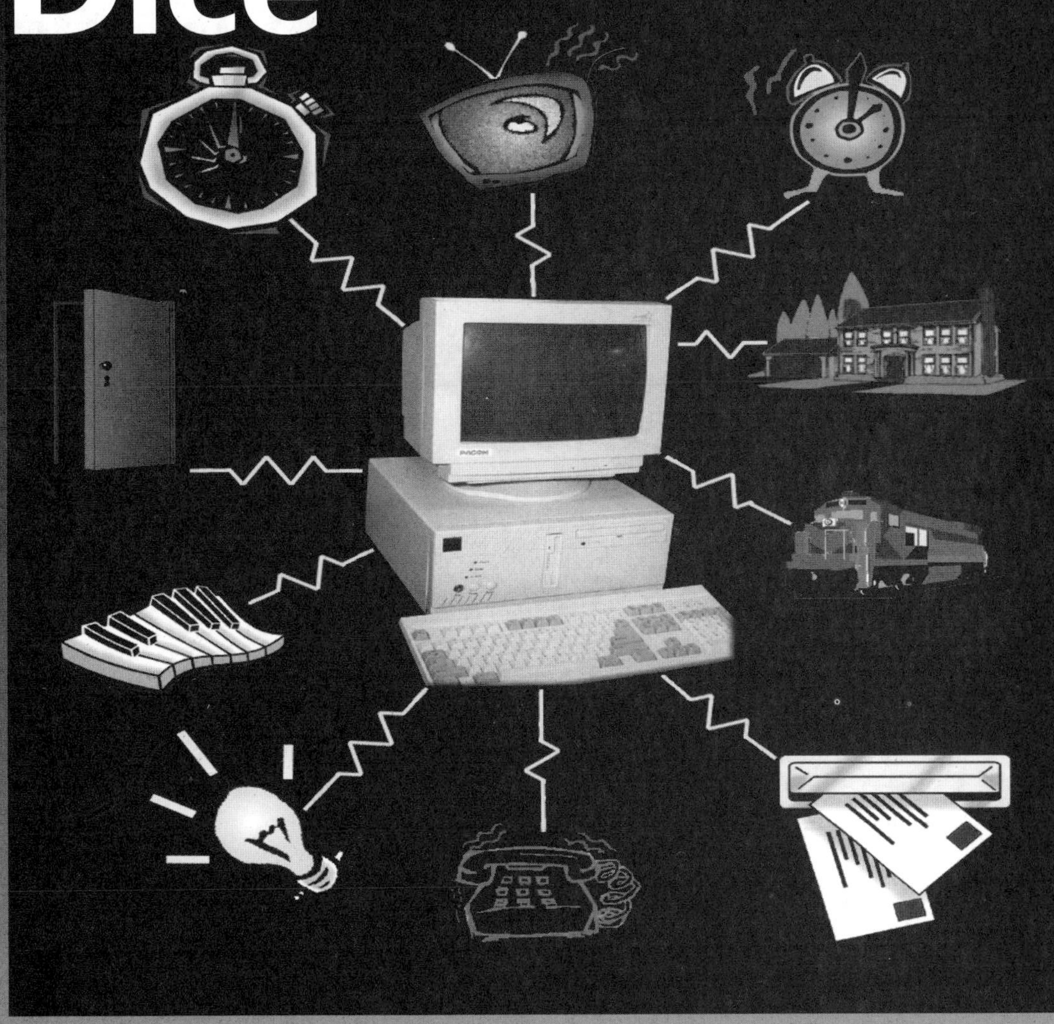

# Chapter 12: Electronic Dice

You're probably familiar with dice from many different games. The circuit of this chapter simulates electronically the function of a die. When used with a program from the companion CD-ROM, you can simulate a die. Light-emitting diodes will be used to show the results on the circuit.

You'll build the circuit on half of a European ("pre-perfed") standard circuit board. You will need the basic circuit from Chapter 9. This circuit doesn't require any special components.

## The Circuit And Its Function

The circuit functions quite simply. After starting the program, you press a button using the mouse to trigger the "roll" of the die. The circuit outputs the result of the roll.

The light-emitting diodes on the circuit are arranged like the dots on a real die so you can easily read the result of the roll.

## Building The Circuit

Prepare half a standard circuit board for building the circuit. You will also need seven light-emitting diodes and a 21-pin edge connector.

Solder the components onto the board as shown in the following component layout.

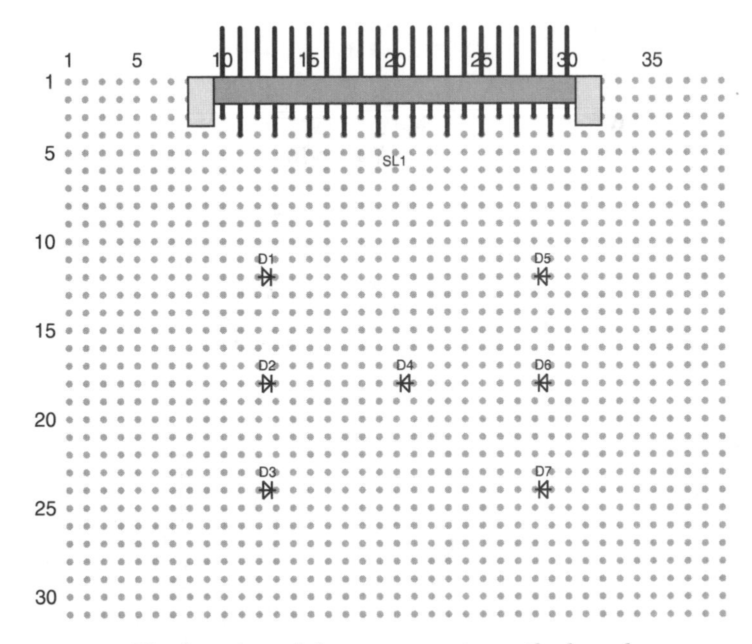


*The location of the components on the board*

You can quickly build the circuit. Pay attention to the polarity when working with the light-emitting diodes; it won't always have the same direction.

## Parts list

Naturally, our chapter wouldn't be complete without a parts list for the circuit.

| Parts List | | |
|---|---|---|
| Label | Name | Component type |
| LED1 - LED7 | Light-emitting diode 5 mm, green | Light-emitting diode |
| SL1 | Pin strip 21pin | Edgeboard connection |

## The Foil Side Of The Board

This circuit only requires a few connections on the foil side of the board. Refer to the following drawing for the necessary connections.

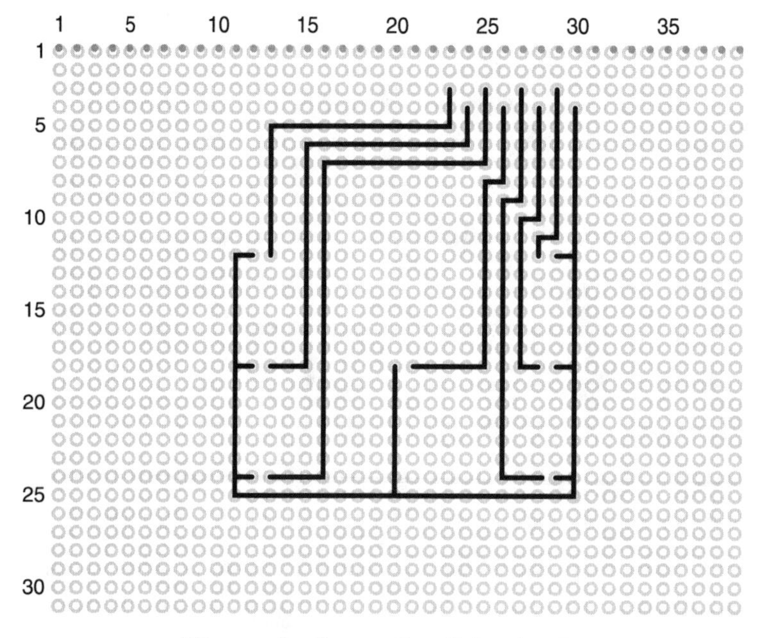


*The conducting paths of the circuit*

After you finish soldering the components and the conducting paths, the finished circuit will look like the following illustration.

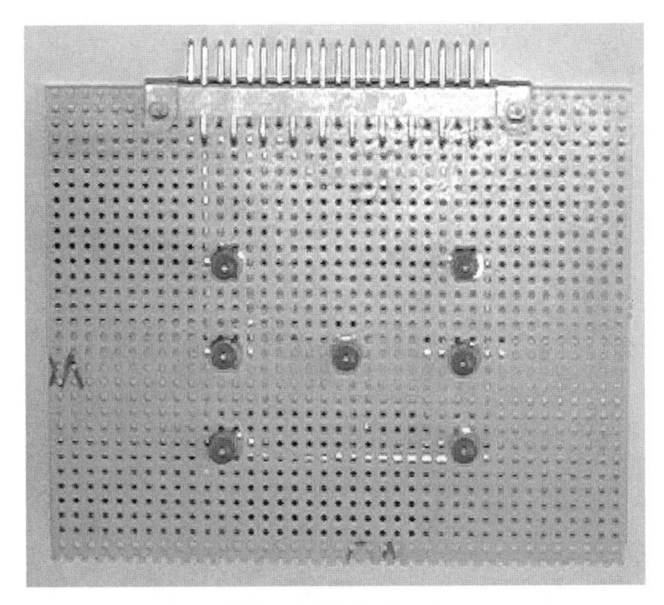

*The finished circuit*

## Using The Circuit With The Companion CD-ROM

The program for this circuit is located in its own folder called *Chap_12* on the companion CD-ROM. It also has a subfolder called *Windows* and two subfolders called *Install* and *Source*.

Again, you'll need to run the *Setup.exe* file to install the program. After installation is complete, you can start the program. The window that appears after startup doesn't contain many elements, making it easy to understand the functions.

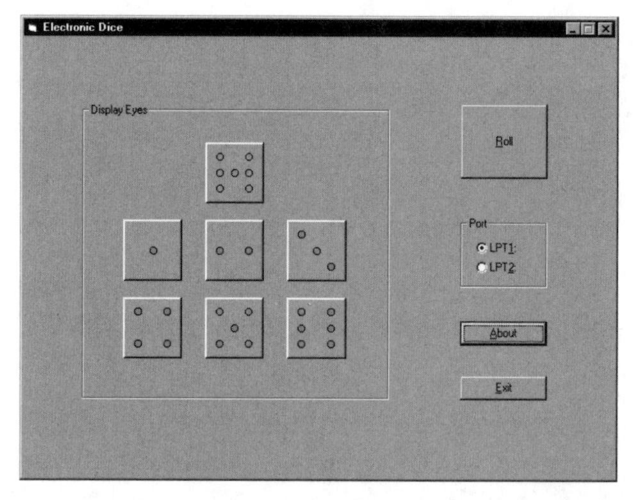


*Running the program*

In the right area of the window, you see the option for determining the port to which the circuit is connected.

To roll the die, click on the *Roll* button. The program then calculates the result of the roll. After that, the circuit displays the output, so that you can read the results from the light-emitting diodes that light up. During calculation, or during the roll, first all the light-emitting diodes are turned on, then off. After that you see the display of the "rolled" eyes on the die.

To display a specific number of eyes on the die, use the buttons in the *Display Eyes* area. Depending on which button you click in this area, the appropriate information is transmitted to the circuit and displayed immediately.

## Using The Circuit

Only a few steps are required to operate the die.

1. Connect the circuit to the parallel port of your computer using a printer cable.

2. Plug the die circuit into the basic circuit and supply the basic circuit with the necessary power. Pay attention to the polarity of the power supply.

3. Next, start up the program. The electronic die is ready to use.

4. Click on the *Roll* button to roll the die. The circuit will output the result of the roll for you.

# Chapter 13:

# The Electronic Thermometer

# Chapter 13: The Electronic Thermometer

The circuit in this chapter measures the current temperature where the sensor is located. The circuit transmits the reading to the program running on the computer, which then displays the temperature reading.

You can use the circuit to measure the room temperature or the outside temperature. Since the program stores both the highest and the lowest reading, you can recognize the maximum values over a specific time period.

You will build the circuit on half of a European ("pre-perfed") standard circuit board and connect it to the parallel port of your computer.

## The Circuit And Its Function

This circuit uses a temperature sensitive resistor to measure the temperature. So the sensor can be placed almost anywhere, we won't solder it directly to the board. Instead, we'll connect it to the circuit using a cable.

This circuit uses a device called an ADC (or Analog-to-Digital Converter). It's used to convert an analog signal into a digital value. The circuit uses an ADC to convert the temperature sensor's reading into a number, which is sent to the computer over the parallel port. The program analyzes this number and displays the appropriate temperature.

This circuit requires an external power supply, so we'll use the same power supply from earlier circuits.

## Building The Circuit

The circuit doesn't have many components so we can easily build it on half of a European ("pre-perfed") standard circuit board. In addition, you will need a flexible two-conductor cable so you can move the temperature sensor to where you want to measure the temperature without moving the circuit.

This circuit also requires a connection to the printer cable, so you will need to get a flat ribbon cable with the Centronics plug for this circuit. Once you have prepared half of a standard circuit board, solder the components as shown in the following drawing.

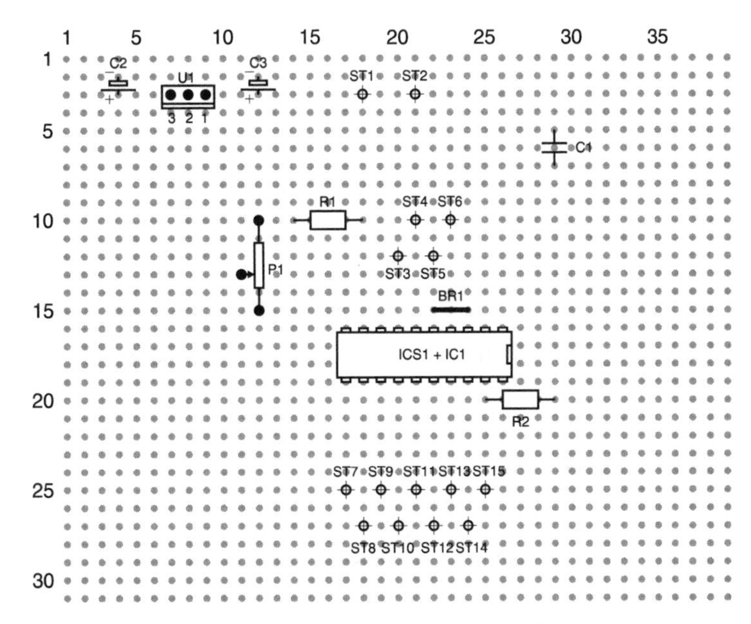


*The temperature measurement circuit*

### Parts list

Check this parts list for the components you will need.

| Parts List | | |
|---|---|---|
| Label | Name | Component type |
| BR1 | Conducting path bridge | Plastic-coated copper wire |
| R1 | 6.2 kΩ | Resistor |
| R2 | 10 kΩ | Resistor |
| P1 | 2 kΩ | Spindle potentiometer |
| C1 | 150 pF | Condenser |
| C2,3 | 22 µF | Condenser (Elko) |
| IC1 | ADC0804 | A/D converter |
| ICS1 | 20 pin IC socket | IC socket |
| U1 | 7805 | Voltage regulator |
| ST1 - ST15 | Edge connector (female) | Pin connection |

### Building the circuit

Note that the potentiometer used in the circuit can be set using a worm drive. This is important since you cannot set a "normal" potentiometer precisely enough.

There are no special features to consider when you build the circuit. You can solder all the components without problems according to the drawing of the circuit.

## The Foil Side Of The Board

You can get the required connections of the components from the following diagram.

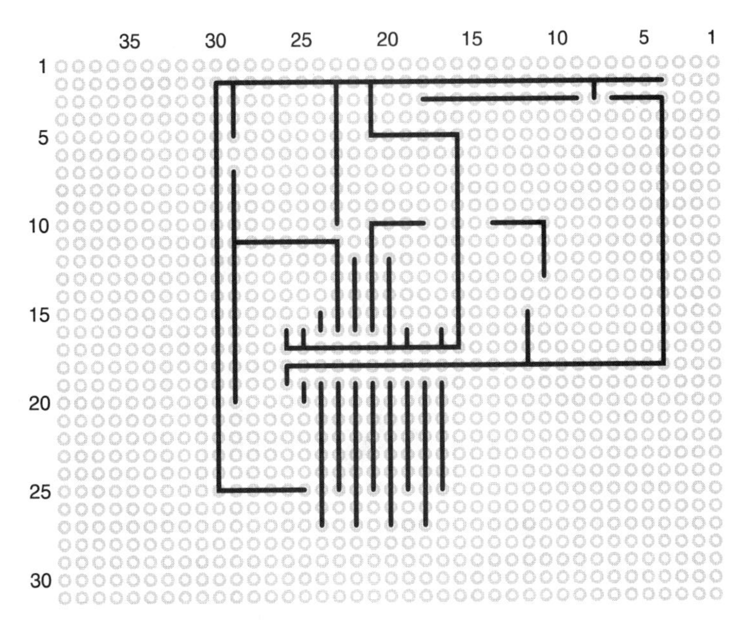


*The foil side of the circuit*

Since we use common components on the board, follow the normal procedure for creating the connections.

## Additional Work On The Circuit

When you finish building the board, several tasks are still necessary before the circuit will work.

First, you need to furnish the circuit with an additional power supply. Use the external power supply that we used for the previous circuits. Using the adapter cable, you can supply the circuit with 9 volts of additional power.

1. For the additional power supply, plug the two male edge connectors into the female edge connectors ST1 and ST2. Remember to plug the positive wire into ST1 and the ground wire into ST2. The voltage regulator on the

board reduces the 9 volts to the required 5 volts. Don't plug the power supply into the socket after making the connection—we still have some other preparations, yet.

Next you need to make the necessary connections from the circuit to the Centronics plug.

2. So the computer can receive the values from the circuit, the flat ribbon cable of the Centronics plug has to be connected to the female edge connectors of the circuit. Look at the following table to determine which wire of the cable connects to which female edge connector.

| ST7 | Wire 21 | | ST12 | Wire 1 |
|---|---|---|---|---|
| ST8 | Wire 19 | | ST13 | Wire 26 |
| ST9 | Wire 23 | | ST14 | Wire 36 |
| ST10 | Wire 25 | | ST15 | Wire 2 |
| ST11 | Wire 28 | | | |

Before turning your attention to the temperature sensor, hook up a simple button between edge connectors ST5 and ST6. The button has to be a locking push button switch, i.e., when you press the button switch, the contact must be closed. You need this button switch to give the circuit a starting impulse.

To connect the button switch, prepare a piece of two-conductor stranded wire about 4 inches long and solder both wires of a male edge connector on one end. On the other end of the stranded wire, connect the two wires with the locking contacts of the button switch. All you have to do then is plug the two male edge connectors into female edge connectors ST5 and ST6. Polarity is unimportant.

Finally, the temperature sensor has to be connected to the circuit. To do this, perform the following steps.

1. Prepare a two-conductor cable (normal strand) of the length required to make a connection from the location of the sensor to the circuit. For example, to measure the outside temperature from your window, you will need a cable that reaches from the window to your computer.

2. On one end of the cable, solder two male edge connectors and plug them into female edge connectors ST3 and ST4 of the circuit. It doesn't matter which cable you plug into which female edge connector, since there is no polarity.

3. Solder two more male edge connectors at the other end of the cable. These will receive the thermal resistor. By using the male edge connectors we are creating a connection that can be broken if necessary. We need a separable connection here because the circuit will have to be set when we start operating it to ensure that we get accurate temperature readings.

4. After soldering the male edge connectors, plug the two pins of the thermal resistor into the edge connectors. The sequence doesn't matter. Make sure that the two pins don't come into contact with each other. You can easily achieve this by wrapping tape around the connectors.

After finishing these tasks, the circuit is ready to be used for the first time.

*The ready-to-use circuit*

## Using The Circuit With The Companion CD-ROM

Before you set the circuit, you need to install its software. The software is located on the companion CD-ROM in the *chap_13* directory. The source files are in a subdirectory of this directory.

Use the setup program to install the software and then start up the software.

To display the current readings on the screen, click on the *Measure Temperature* button. The program then reads the sensor value at regular intervals and displays the current temperature.

In addition, the program remembers the highest and lowest temperature readings. These are also displayed, but you can set them back to zero by hitting the Reset buttons, so that you can determine the maximum values at any time.

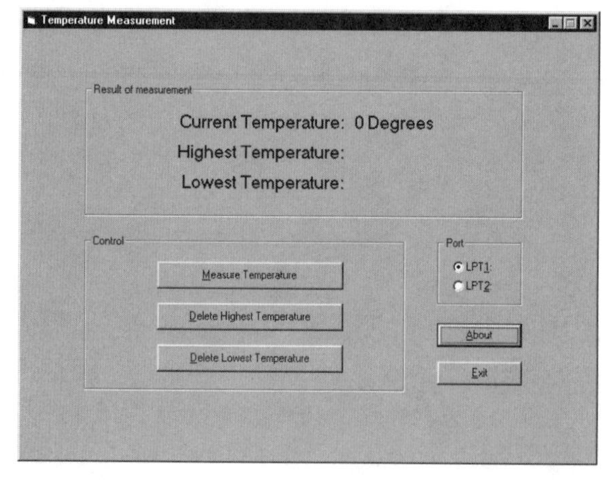


*The program for temperature measurement*

Since the circuit isn't set up yet, the value after the first start won't be correct. You'll have to calibrate the circuit first.

## Using The Circuit

You have completed the necessary work for using the circuit. Now follow the operating procedure described below.

1. Plug the printer cable into the Centronics plug to establish the connection to the computer.
2. Make sure that the sensor is connected to the circuit by the cable.
3. Connect the additional power supply to the circuit, if you haven't already done so, and switch on the power supply.
4. Press the button, so that the circuit gets the starting impulse.
5. Start up the software.
6. Now calibrate the circuit: Disconnect the sensor from the connectors of the cable, and plug a 2-kW resistor in place of the sensor. Now look at the current temperature reading on the monitor. If this value is at approximately 25° Celsius, then the circuit is set up correctly. However, this probably won't be the case.

   If it doesn't read 25°, use a small screwdriver to turn the spindle of the potentiometer. If the value approaches 25° C, keep turning in this direction until you reach the value. If you are moving away from the value, turn the spindle in the other direction until you reach 25.
7. Once the value is set, remove the resistor and replace the temperature sensor. The circuit is now calibrated. Since the temperature sensor has a resistance value of 2 kW at 25° Celsius, the values transmitted now will be correct.

Now you can move the sensor to wherever you wish to measure the temperature.

**Important Tip**

We cannot guarantee the results of these readings since we used low-cost components to build the circuit. The A/D converter has a very low resolution. As a result, the readings are not very accurate. Also, the measuring range of about -50° to about +150° can give large differences in measurement. The circuit cannot measure individual degree jumps. The difference between two readings amounts to at least 3.3°. In other words, don't use this circuit for precision measurements. Please remember this when measuring with this circuit.

# Chapter 14:

# The Electronic Scale

# Chapter 14: The Electronic Scale

As you've probably assumed from the title, the circuit in this chapter weighs objects. It uses a pressure-sensitive electronic component along with the other elements of the circuit to determine the weight of an object within a specific weight range. The value determined by the circuit is then transmitted to the computer. There a program from the companion CD-ROM evaluates the data and displays the corresponding weight to you.

You can build the entire circuit on half of a standard European circuit board. The board is designed to transfer data to the computer through the parallel port.

## The Circuit And Its Function

A pressure-sensitive resistor in this circuit determines the weight of an object. Since the pressure sensor is not soldered directly to the board (it's connected using a cable), you can move the sensor anywhere the cable can reach.

The analog value determined by the pressure sensor is converted into a number that the computer can understand by an ADC within the circuit. This number is then transmitted to the computer through the parallel port. The software evaluates this number and computes the corresponding weight.

We talked about how an ADC operates in Chapter 13 so if you have any questions about this component, please refer to that chapter.

Because the circuit works with an external power supply, make certain to have the external power supply ready after building the board.

## Building The Circuit

The circuit requires only a few components so it will fit on half of a standard European ("pre-perfed") circuit board. You will need a flexible two-conductor cable to which the pressure sensor is attached.

This circuit, like the one from Chapter 13, requires the flat ribbon cable with the Centronics plug. This is because this circuit also transmits data through the parallel port to the computer. Therefore, have the cable ready for this purpose.

Refer to the following drawing for the layout of the components. Then solder them on the board.

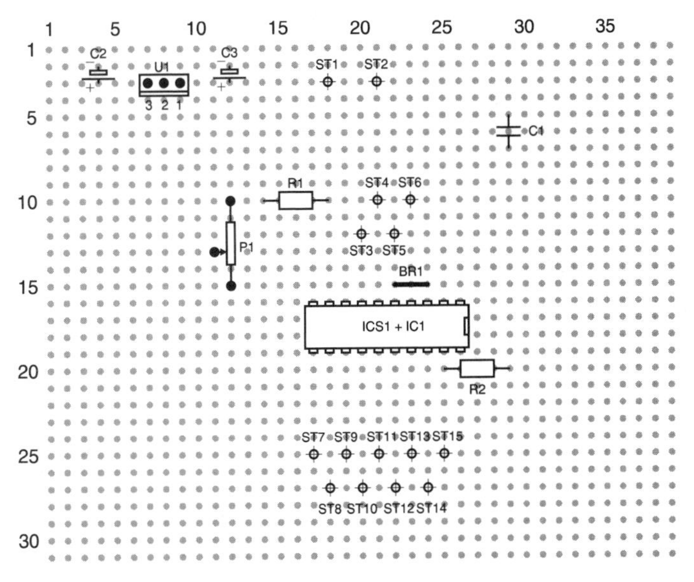


*The weight measurement circuit*

### Parts list

Refer to the parts list in the following table to determine which components are used in this chapter.

| Parts List | | |
|---|---|---|
| Label | Name | Component type |
| BR1 | Conducting path bridge | Plastic-coated copper wire |
| R1 | 100 kΩ | Resistor |
| R2 | 10 kΩ | Resistor |
| P1 | 47 kΩ | Spindle potentiometer |
| C1 | 150 pF | Condenser |
| C2,3 | 22 µF | Condenser (Elko) |
| IC1 | ADC0804 | A/D converter |
| ICS1 | 20 pin IC socket | IC socket |
| U1 | 7805 | Voltage regulator |
| ST1 - ST15 | Edge connector (female) | Pin connection |
| external connection | FSR 151 | Pressure sensor |

### Building the circuit

The potentiometer used in this circuit is also set using a worm-drive. Only with this type of setting option is there a guarantee that the required calibration can be performed precisely.

## The Foil Side Of The Board

Refer to the following diagram for information on the conducting paths.

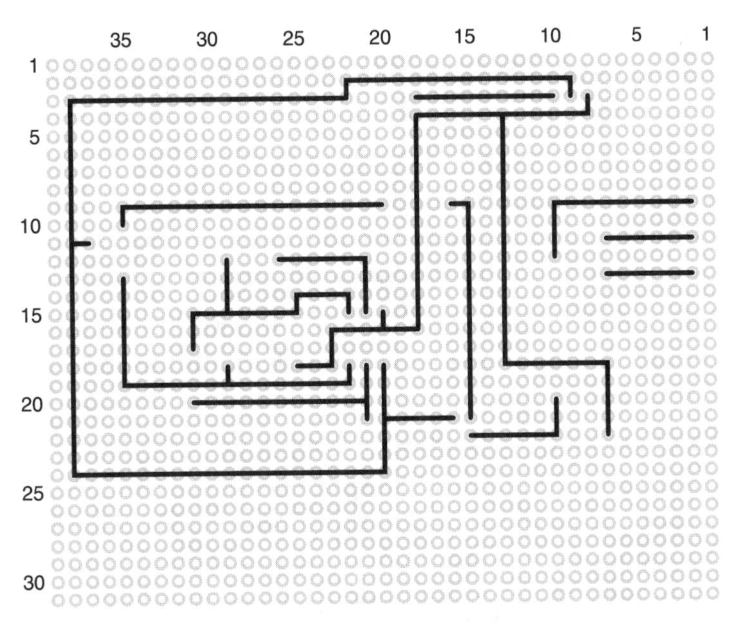


*The foil side of the electronic scale*

Solder the conducting paths carefully. Since there are no unusual components in this circuit, the soldering shouldn't cause any difficulty.

## Additional Work On The Circuit

The finished circuit cannot do its work without the pressure sensor. Therefore, we'll connect it to the circuit.

However, before we do this, we'll need to provide the circuit with additional power. We'll use the external power supply that we used with previous circuits. Use the adapter cable to furnish the circuit with 9 volts of additional power.

The additional power supply plugs into the two female edge connectors ST1 and ST2. Plug the positive wire into ST1 and plug the ground wire into ST2. Now the circuit has an external power supply. However, don't plug it into the socket yet. A few other preparations are still necessary.

Besides the additional power supply, you'll also have to connect the flat ribbon cable with the Centronics plug to the circuit board. Keep the following in mind when you do this:

Use the same cable to connect the circuit to the parallel port of the computer that you used in Chapter 13. If you haven't yet built the circuit in Chapter 13, refer to this chapter to find out how to create the cable. It's important that you make the cable exactly as described in Chapter 13, since otherwise the individual wires might not be connected with the correct female edge connectors.

The following table will help you determine which wire of the cable goes with which edge connector.

| | | | | |
|---|---|---|---|---|
| ST7 | Wire 21 | | ST12 | Wire 1 |
| ST8 | Wire 19 | | ST13 | Wire 26 |
| ST9 | Wire 23 | | ST14 | Wire 36 |
| ST10 | Wire 25 | | ST15 | Wire 2 |
| ST11 | Wire 28 | | | |

You'll also have to attach a push button switch to female edge connectors ST5 and ST6, as you did for the circuit from Chapter 13. This button supplies the starting signal for the circuit. Follow the steps in Chapter 13 to build the cable. It's also connected the same way.

We still have to deal with the pressure sensor. Without this component, the circuit will not be able to measure weight.

The pressure sensor is a component labeled FSR 151. As you can see in the following illustration, it resembles a flattened spoon.

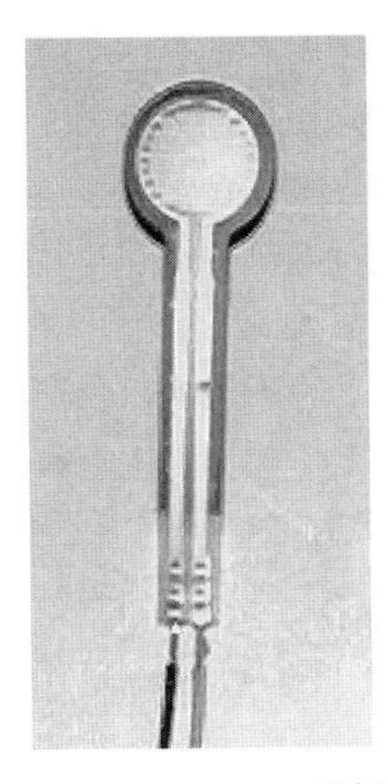

*The pressure sensor FSR 151*

The plate-shaped end is the pressure-sensitive area of the component. This is where you will place the object to be weighed. The other end has two little pins where the measuring wire is soldered.

Follow these instructions to connect the pressure sensor to the circuit:

1. First, you will need a two-conductor cable (normal stranded) long enough to create a connection from the measuring area to the circuit.
2. Prepare the cable for use. On one end of the cable, solder two male edge connectors and plug them into female edge connectors ST3 and ST4 of the circuit. Since the sensor doesn't have any polarity, you are free to choose the arrangement.
3. Strip the other end of the cable so you can solder a resistor or the pressure sensor to it. However, don't solder either one to the cable yet.

After completing these tasks, the circuit is ready to be put into operation for the first time and to be calibrated.

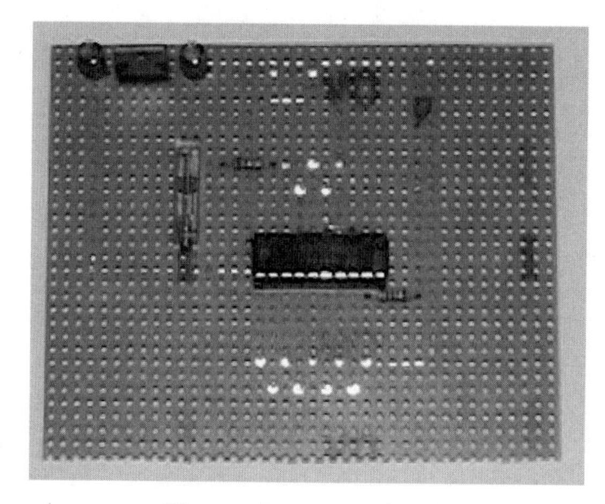

*The ready-to-use circuit*

## Further preparations

You've completed the circuit to the point that it can accomplish its task. However, you still need something to use for measuring the weight.

You learned that the pressure sensor has a round area on which the weight to be measured is placed. This area has a diameter of approximately 3/4 inch. It would be fairly difficult to place the object you wish to measure on that area. What you need to do is build a device that lets you place objects on it, and at the same time, transfers the weight to the appropriate area of the pressure sensor.

If possible, it should be a plate on which you can place the object to be weighed and on whose bottom side a cylinder of approximately. 3/4 inch diameter with a low height is attached.

When creating this "plate," remember to give it a guide so that it doesn't always tip over if you place something on it that is not exactly in the middle of the plate.

**Important Tip**

Also, since you have a total measuring range of 10 g to 10 kg (.026 pounds to 26.8 pounds), make sure the plate isn't too heavy. For example, if the plate itself weights 5 kg (13.4 pounds), you can only measure weights up to 5 kg (13.4 pounds).

## Using The Circuit With The Companion CD-ROM

As we mentioned, the companion CD-ROM includes a program to evaluate the measurement from the circuit. You'll find this program in the *chap_14* directory. Run the *Proj_14.exe* file to perform the installation. You will find the source files for the software in a subdirectory.

Install and then start the program.

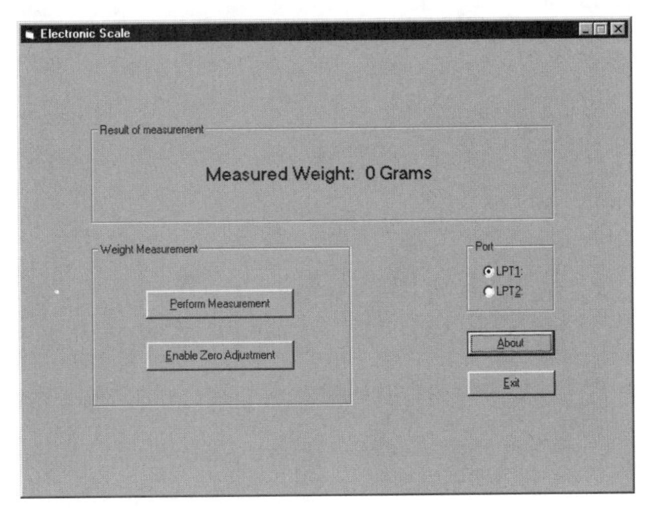


*The program for weighing objects*

The program has some buttons and options for your use.

You're already familiar with the option for which port you use (LPT1 or LPT2). Enable the appropriate option to have the software look for the correct port.

The *Zero Adjustment* button has to do with the measuring plate we were just discussing. Once you have calibrated the circuit as specified further below, you can use this button to toggle off the weight of the measuring plate. To do this, place the measuring plate on the pressure sensor and then click the button. The software program then registers the weight of the measuring plate and no longer considers this weight when making subsequent measurements. You get the weight reading for the object you are measuring without the weight of the measuring plate being included in the reading.

Press the *Perform Measurement* button to transfer the measured value of the circuit to the software program. Place the object to be measured on the measuring fixture and then click on this button. The software program then displays the weight. The display won't change even if you remove the weight to be measured or place something on it until you click *Perform Measurement* again.

Since the circuit hasn't yet been set, the first time you start the circuit up the value will not be correct. You will have to calibrate the circuit first. We'll tell you how to do this in the next section.

## Using The Circuit

The following steps give you instructions on operating the circuit as well as the procedure for calibrating the circuit.

1. Solder a 2   ohm resistor to the free end of the measuring wire, so that both wires are connected to each other by the resistor.
2. Using the printer cable, connect the circuit to the computer.
3. Connect the additional power supply to the circuit and switch on the power supply.
4. Start the circuit by pressing the button.

5. Start up the software. Set the correct port and click on the *Perform Measurement* button. The program then displays a reading.

6. Calibrate the circuit by turning the small screw on the spindle potentiometer to the left or to the right. After clicking on the button a second time, the displayed reading within the software will vary correspondingly.

**Important Tip**

Set the potentiometer so that the software displays a value of approximately 10 kg (26.8 pounds). This calibrates the circuit.

7. Once the value has been set, remove the resistor and solder the pressure sensor to the measuring wire in its place. The circuit is then ready to use.

8. Now place the measuring plate on the pressure sensor and click the *Zero Adjustment* button. Pressing this button causes the software to register the measuring plate, and you can begin measuring the weight of objects.

If you determine that the measurements aren't quite correct, you can calibrate the circuit using an exact weight. To do this, place the weight on the pressure sensor (without the measuring plate) and turn the screw on the potentiometer until the software displays the appropriate weight.

However, the circuit cannot perform precision measurements, since you are working with "normal" components that have relatively high tolerance values. For absolutely correct measurements, you would need to use higher quality components and build a more complicated, expensive circuit. This would ultimately drive the expenses and work involved in building the circuit beyond our "ceiling of limitation."

**Important Tip**

Keep in mind that you won't be able to measure subtle weight differences, since the resolution of the measuring circuit is not very great. Remember too, that the measurements in some ranges are subject to rather high tolerances, especially in the ranges of extreme values close to 10 kg (26.8 pounds) or in very small ranges under 1 kg (2.67 pounds).

# Chapter 15:

# The Regulated Light Control System

# Chapter 15: The Regulated Light Control System

The circuit in this chapter reacts to the level of brightness. In other words, it monitors the ambient light and responds according to the setting(s) you define. For example, you can switch on a light when a specific level of darkness is reached.

You can build the circuit on half of a standard European ("pre-perfed") circuit board and operate it through the parallel port of your computer.

## The Circuit And Its Function

The circuit measures the quantity of light using a brightness-sensitive resistor called a *photoresistor*. We'll connect the photoresistor to the circuit using a two-conductor cable so you can move the sensor to different locations.

The value generated by the photoresistor is evaluated within the circuit and transmitted to the computer through the parallel port. The program on the companion CD-ROM accepts this value and, depending on the program options, you can control a relay on the circuit. The locking or opening contacts of the relay are accessible on the board through a (connecting) terminal, so that you can connect, for example, the plug of a lamp there.

This circuit requires an external power supply.

## Building The Circuit

This is a very easy circuit to create although some additional work is necessary. This is especially true for the photoresistor.

This circuit also requires the familiar flat ribbon cable with the Centronics plug to transfer the data through the parallel port to the computer. You'll need to get the cable ready for this circuit.

Before tackling the additional work, you'll have to build the circuit. Consult the following drawing for the components and their layout.

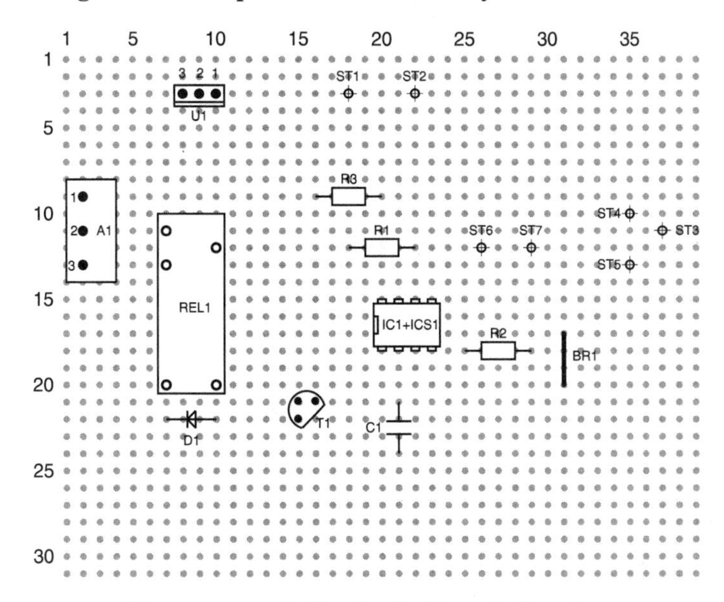


*The components for the light control system*

## Parts list

Refer to the parts list in the following table to determine which components are used in this chapter.

| Parts List | | |
|---|---|---|
| Label | Name | Component type |
| BR1 | Conducting path bridge | Plastic-coated copper wire |
| R1, R2 | 1 kΩ | Resistor |
| R3 | 2.2 kΩ | Resistor |
| T1 | BC546B | Transistor |
| C1 | 0.47 µF | Condenser |
| D1 | 1N4001 | Diode |
| IC1 | NE555 | Timer-IC |
| ICS1 | 8 pin IC socket | IC socket |
| U1 | 7805 | Voltage regulator |
| REL1 | 5-V power relay 1 x UM | Card relay |
| A1 | 3 pin connecting terminal | Connecting terminal |
| ST1 - ST7 | Edge connector (female) | Pin connection |

## Building the circuit

A few explanations are in order for building the circuit and about specific components. When soldering the diode, pay attention to the polarity.

1. Make sure you insert the IC into the socket correctly and that you solder the socket as shown in the drawing.

   We cannot include a detailed description of the *relay* since so many of the relays are made by different manufacturers. Fortunately, they all have very similar construction and similar performance data.

2. The relay must work with a voltage between 5 V and 6 V and have an alternating contact. It must be designed for print assembly on a standard circuit board. The coil resistance of the relay should be about 55 ohms.

   The contacts are usually standardized so that doesn't present a problem. Depending on what you want to control with the circuit, pay attention to the performance data of the relay. It should be able to convert 220 volts. Furthermore, consider the specified maximum contact current of the relay. It may not be exceeded—otherwise the relay could be destroyed.

3. The *connecting terminal* must be designed for circuit board assembly. This is the only type of connecting terminal you can solder properly to the board using the existing pins. Other than that, you may be familiar with the operation of terminals from thermoplastic connectors. Here, too, you can insert a contact wire into an opening and secure it with a small screw. Keep in mind that connectors 1 and 2 represent a normally open contact, while connectors 1 and 3 represent a normally closed contact.

After soldering the components to the board, you can turn your attention to the foil side of the board.

## The Foil Side Of The Board

The components of the board must be connected with each other by the conducting paths shown in the following drawing.

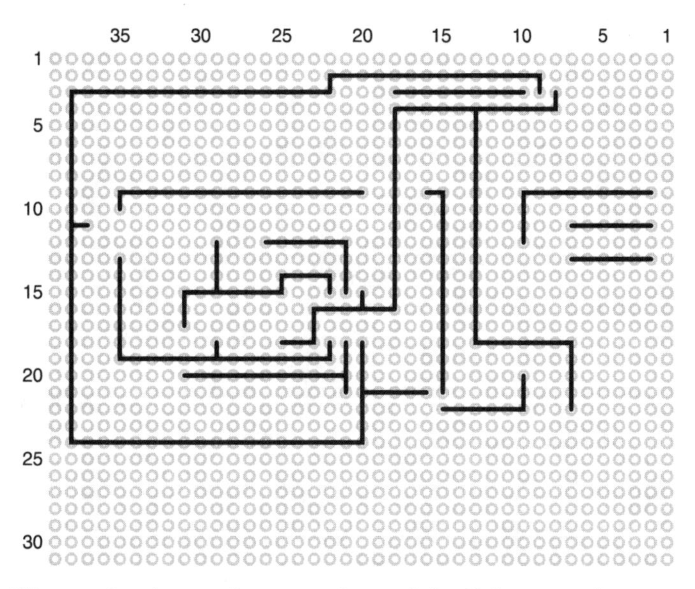


*The conducting path connections of the light control system*

Once the conducting paths are attached to the board, you still need to perform some additional work before the circuit will operate.

## Additional Work On The Circuit

As you know, the circuit employs a photoresistor for light measurement. This component is labeled M9960. The photoresistor changes its value of resistance according to the quantity of light on the component. The circuit uses this varying resistance for the measurement.

The photoresistor is not included on the board. It is to be connected to the circuit by a cable (similar to the pressure sensor in Chapter 14). That allows you to determine independent of the circuit where to measure the light level.

1. Like a normal resistor, the photoresistor has two contacts. As a rule, the contacts consist of two long wires that are attached to the actual sensor. Cut each of the wires to a length of 3/4 inch.

2. Then take a two-conductor cable and strip approximately 1/8th of the cable ends. Then solder the conductors of one cable end to the contacts of the photoresistor. Polarity doesn't play a role. However, make sure the contacts of the photoresistor cannot come into contact with each other. If necessary, insulate the contacts with adhesive tape. Insulate the contacts with adhesive tape or heat shrink tubing.

3. Equip the other end of the cable with male edge connectors. Then plug them into female edge connectors ST6 and ST7. Here again, polarity is not a concern.

4. Attach the additional power supply the circuit requires to the two female edge connectors ST1 and ST2. Here you do have to watch the polarity. The positive wire plugs into female edge connector ST1 and the ground wire plugs into female edge connector ST2.

This circuit also requires the flat ribbon cable with the Centronics plug. To be able to properly exchange data between the circuit and the computer or software, the following connections are necessary.

| ST3 | Wire 2 |
|---|---|
| ST4 | Wire 3 |
| ST5 | Wire 19 |

Once you have attached the photoresistor, the external power supply and the cable from the computer port to the circuit, the circuit is ready to use.

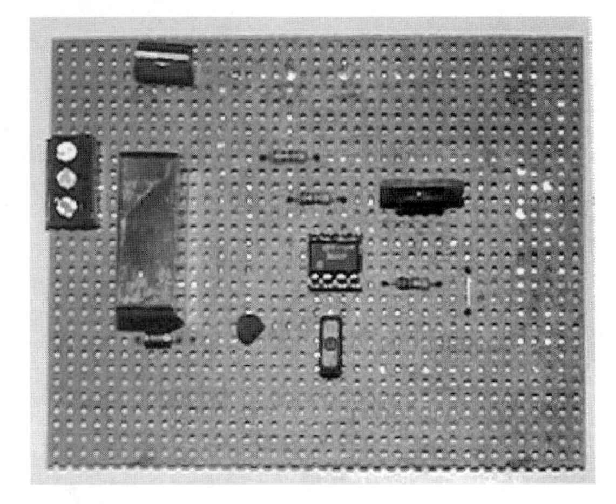

*The finished circuit for measuring light*

## Using The Circuit With The Companion CD-ROM

The circuit works with a program that you'll find on the companion CD-ROM in the *Chap_15* directory. You can install the software by running the *Setup* program.

The *Windows* directory has a subdirectory named *Source*, which contains the source files for the software. Use these files to make any changes to the software.

After you install and run the program, you'll see the following screen:

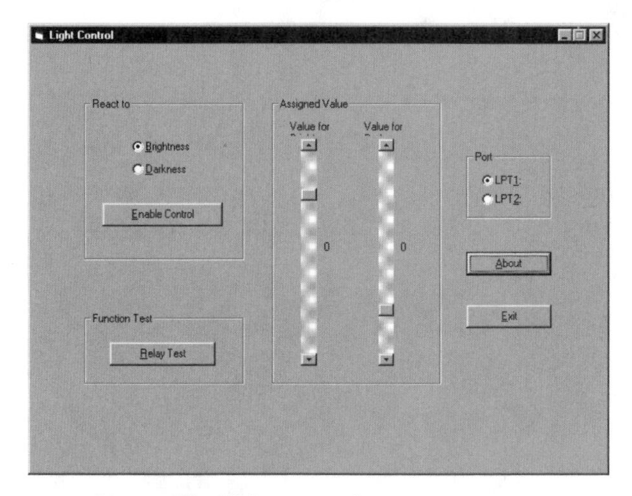


*The light control program*

You're already familiar with the option for setting the port (LPT1 or LPT2). Set this option to indicate which of the two parallel ports you are using.

Before turning your attention to the setting options, take a look at the *Relay Test* button. When you click this button, the relay is tested. The software switches the relay on and then switches it off again after about 2 seconds. This way you can check the function of the relay or the effect of what is being controlled by the relay regardless of the incidence of light and the proper function of the light measurement.

Another important setting option for using the circuit is in the *React To* area, with the options *Brightness* and *Darkness*. Depending on which of the options you enable, the software will respond to the presence of an adjustable brightness or an adjustable darkness. The setting you choose here determines whether the relay will be enabled in brightness or in darkness.

To have a light switched on when a specific level of darkness is attained, enable the *Darkness* option. On the other hand, if you wish to lower the Venetian blinds upon attainment of a specific level of brightness, you would enable the *Brightness* option.

Use the *Value for Brightness* and *Value for Darkness* sliders to specify the value to which the circuit reacts. Since it is not possible with these settings to work with precise lux specifications (lux is a unit of illumination), we provided the sliders. You'll have to experiment to find the right settings.

To do this, operate the circuit as described in the next section and wait until a specific brightness or darkness appears on the photoresistor. Then move the slider in such a way that it activates the relay. Then you can note the value and use it for recurring operations.

You'll need to do some experimenting with the values to learn how to get the circuit to react in the desired manner.

## Using The Circuit

Operating the circuit is relatively easy. You're already familiar with some of these steps, however, be sure to read the "Important Tip" message at the end of this chapter.

1. First, plug the cable of the photoresistor into the appropriate female edge connectors of the circuit and position the sensor in the desired measuring area.

2. Then connect the circuit to the computer. Use the printer cable to make the connection.

3. Connect the external power supply to the circuit and plug the cable of the power supply into the socket.

4. After starting up the software program, set the correct option for the port. Then click on the *Relay Test* button to test the function of the relay.

5. Select the correct option in the *React To* area, depending on whether you want the circuit to react to brightness or darkness.

6. Wait until a specific brightness or darkness is attained, or else cover up or direct light toward the sensor to provide the desired state of light, then move the slider of the appropriate option until the desired reaction occurs. Since this requires a lot of experimentation, be sure to allow yourself plenty of time.

Once you have made the settings and the circuit reacts to a specific brightness or darkness, you can use the circuit for control purposes. However, make certain to read the rest of this chapter; these paragraphs include very important information.

## Be careful with the circuit

Using the terminals on the board allows you to control a source of light or a similar electrical appliance. This is done by using closing or opening contacts through the terminals.

This means that the circuit to be controlled, for example a light source, is closed or interrupted at the terminals and the desired effect (the light goes on or off) is achieved as a result of this. There's usually no problem with this. However, under certain circumstances you'll have to use special caution.

To better understand the importance of the following paragraphs, pick up the circuit and look at the relay and at the terminals, in particular the wire connections of these two components on the rear side of the board. As you can see, the contacts of the relay are connected to the terminals by bare, soldered wire lines. This circumstance requires special caution.

As long as you use the circuit, for example, to control a 12 volt circuit of a model train light system, the circuit layout doesn't present any danger. However, it's a different story when it comes to using the circuit to control a 220 volt circuit. If you wish to use our circuit to switch on a 220 volt lamp, you cannot simply do it. There's a possibility that you will unintentionally touch the bare wires on the board. This could be hazardous to your health.

Even if you don't come directly into contact with the wires, there could be a flow of current through the object on which the circuit is lying. You definitely don't want this to happen. Pay careful attention to the following "Important Tip."

**Important Tip**

If you intend to control high voltage through the terminals of the circuit, find a good electronics shop and have a non-conducting case placed around the circuit. This will protect you from direct contact with the conducting paths. If you use our circuit to control alternating voltages, you will want to do this even with lower voltages.

# Chapter 16:

# The Exact Measuring System

# Chapter 16: The Exact Measuring System

The measuring system this circuit creates allows you to pass a specific quantity of liquid through a *flow meter*. By using a relay, you could control a pump, for example, making it possible to dispense a specific quantity of liquid with exact measurements.

One conceivable use would be supplying a window box with a specific amount of water, or changing the water of an aquarium, where a precisely specified quantity of water could be pumped out and then replaced by an equal amount of fresh water.

The basic circuit from Chapter 9 manages data exchange through the parallel port. In addition, the basic circuit furnishes the power supply. Get these two items ready.

Also, you will need half of a standard European ("pre-perfed") circuit board to build the circuit.

## The Circuit And Its Function

We've kept the structure and the operation of the circuit relatively simple. However, using the circuit with the corresponding program from the companion CD-ROM is not quite as simple, since the program has a great deal to achieve in this case. We'll tell you more about that later.

The circuit lets you connect a flow meter that gives off impulses as soon as liquid is flowing through it. The program gathers and processes these impulses. A relay, controlled by the program, gives you the option of additional control processes through a closing contact.

You'll have a better idea of how the circuit functions in the "Using The Circuit With The Companion CD-ROM" section.

## Building The Circuit

The circuit consists of components with which you are familiar, so you'll be able to build the circuit easily. Once you finish building the circuit, you can quickly begin using it.

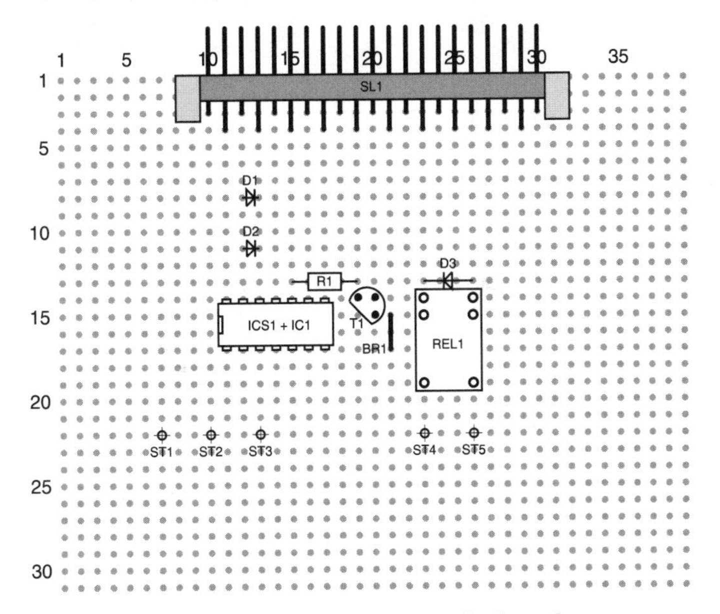


*The component layout on the board*

## Parts list

Refer to the parts list in the following table to determine which components are used in this chapter.

| Parts List | | |
|---|---|---|
| Label | Name | Component type |
| BR1 | Conducting path bridge | Plastic-coated copper wire |
| ST1 - ST5 | Edge connector (male) | Pin connection |
| R1 | 2.2 kΩ | Resistor |
| T1 | BC546 | Transistor |
| D1 | Light-emitting diode 5 mm (red) | Light-emitting diode |
| D2 | Light-emitting diode 5 mm (green) | Light-emitting diode |
| D3 | 1N4007 | Diode |
| IC1 | SN7404 | Inverter |
| ICS1 | 14-pin IC socket | IC socket |
| REL1 | Small relay 5 V, 1-pin UM | Relay |
| SL1 | Pin strip | Pin connection |

When building the circuit, make sure the transistors and diodes aren't mounted the wrong way. Also, the IC and the IC socket must be soldered in correctly.

# The Foil Side Of The Board

Refer to the following diagram for information on the conducting paths.

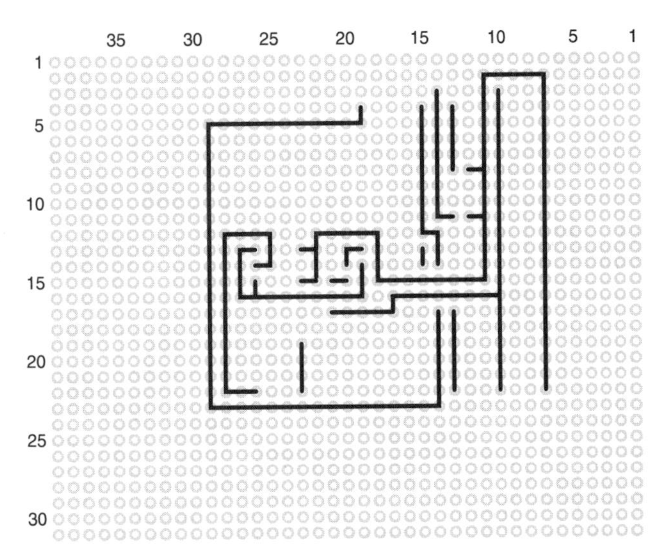


*These conducting paths connect the components*

When you finish building the circuit, it should look like this::

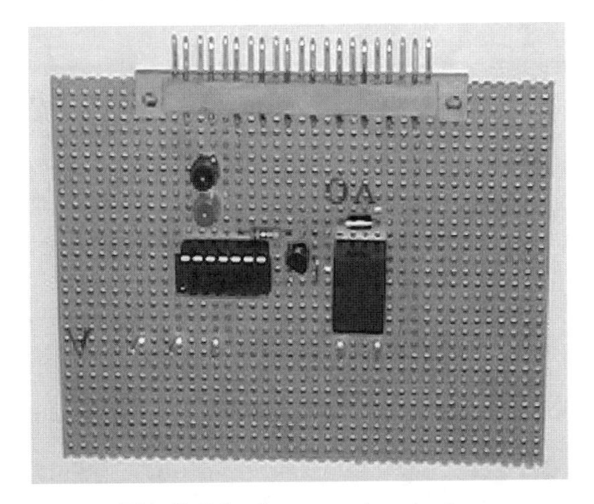

*The finished measuring system*

Some more work is necessary before the circuit can operate (see the next section).

## Additional Work On The Circuit

As we mentioned, the circuit requires a *flow meter* as a sensor. You can get this at an electronics shop. There are many brands of flow meters and, some are quite expensive The following illustration is an example of a reasonably priced flow meter.

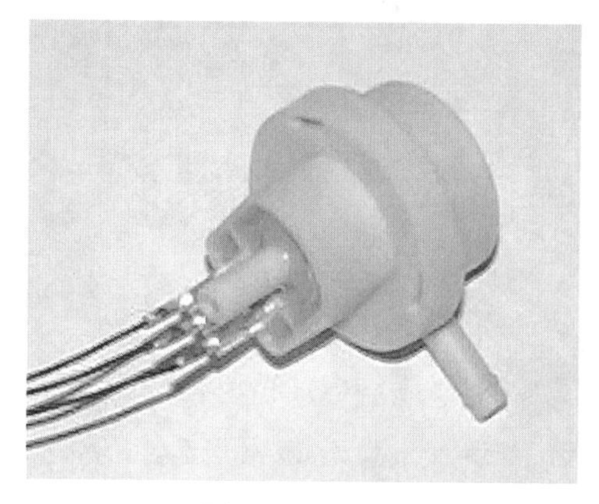

*The flow meter*

This flow meter must be supplied with power using two connectors and then delivers the impulses at a third connector when liquid is flowing through it. The circuit is designed accordingly. For example, female edge connector ST1 supplies five (5) volts and female edge connector ST2 furnishes the ground. The sensor's signals are sent to female edge connector ST3.

Solder the male edge connectors to an appropriate cable so you can make the correct connection from the circuit to the flow meter. The component shown in the illustration above requires positive on the right pin and ground on the middle pin, if you view it from the front. The impulses are delivered to the left pin.

If you decide to use a different flow meter, make certain it can supply the appropriate impulses so the circuit can evaluate them. It must also operate on 5VDC.

The circuit is finished when you connect the sensor. It's your option on how you link the flow meter to the system using the small hoses, whose fluid balance you want to influence.

Make absolute certain that you only work with as much pressure within the hoses as the flow meter can bear. In addition, use only fluids that will not harm the sensor. For more information, read the documentation that comes with the flow meter.

## Using The Circuit With The Companion CD-ROM

The program for the measuring system has an important responsibility. However, before you can use it, you must install it. The appropriate files are on the companion CD-ROM in the *Chap_16* directory. The *Setup* program guides you through the installation.

The source code files for the program are located in the *Source* subdirectory, in case you want to change the circuit's function.

When you start the program, the following screen appears:

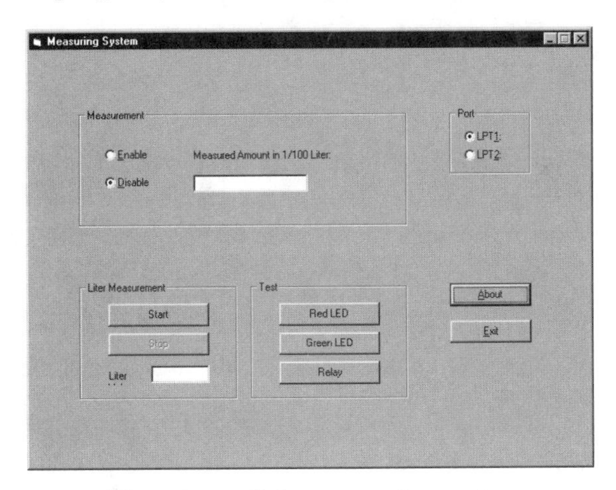


*The screen of the measuring system*

Since the circuit uses the parallel port, the option for LPT1 or LPT2 is included. Choose the port through which you will be running the measuring system.

The *Test* area provides you with three buttons. You can test whether the light-emitting diodes and the relay are working properly. When you click on one of the buttons, the appropriate component is switched on for approximately 1 second. Then it automatically switches off again.

When you have successfully performed the tests, your next step is to perform a *Liter Measurement*. This adapts the circuit to your flow meter. Before you can run a liter of fluid through the flow meter, click on *Start*. Then run the fluid through and click on *Stop*. Finally, the program outputs a number in the *Liter Value* input line. This number is the indicator for the circuit. Take note of the number, so that you can enter it there the next time you start the program. Then you won't have to run the liter test again. However, be sure you don't enter a nonsense number. The program won't run a plausibility check.

After performing the liter measurement, you can enter the *Measured Amount*. Enter this as a whole number in 1/100 of a liter. For example, if you enter the number 50, it means you wish to measure half a liter of liquid.

After specifying the amount to be measured, click the *enable* option. As a result, the following events occur: The red light-emitting diode goes out and the green light-emitting diode lights up. At the same time, the relay starts up and closes the ST4/ST5 contact. You could use this, for example, to activate a pump, which then pumps liquid through the flow meter.

The program now checks the impulses continuously, and thus checks the amount of liquid flowing through the flow meter. Once the program registers that the specified amount has run through, it switches off the relay, thus also switching off a connected pump. The light-emitting diodes also change their states. The circuit is then back in the starting position, ready for a new measurement.

To interrupt a measuring cycle, click on the *disable* option. The circuit immediately cancels the measuring cycle.

When entering the measured amount, be sure to enter a reasonable value. The program doesn't perform a plausibility check. Nonsensical entries can result in a nonsense measurement.

## Using The Circuit

The following steps give you instructions on operating the circuit.

1. Plug the board into the basic circuit from Chapter 9.
2. Connect the flat ribbon cable to the parallel port of the computer.
3. Make certain that the additional power supply is connected to the basic circuit and that the power supply is plugged into the socket.
4. Connect the output contact of the relay with the component you wish to control, if you are doing this.
5. Place the flow meter so that the liquid to be measured can flow through it.
6. Start the program, and set the correct option for the port you are using.
7. Test the function of the relay and the light-emitting diodes by clicking on the appropriate buttons in the *Test* area.
8. Perform the liter measurement, and then enter the result in the input line.
9. Enter the measured amount. Make sure you specify the amount in hundredths of a liter.
10. Click on *enable* to begin the measuring cycle.

Once you have successfully operated the circuit, you can use the circuit to dispense specific liquid quantities repeatedly.

# Chapter 17:

# The Programmable LED Matrix

# Chapter 17: The Programmable LED Matrix

The circuit in this chapter uses a light-emitting diode matrix to display freely definable characters. The matrix consists of 35 light-emitting diodes that are arranged in five columns and seven rows. Don't worry about having to solder and wire 35 light-emitting diodes—you can buy the matrix in one unit (we'll tell you more about that later).

This circuit is another add-on circuit used with the basic circuit you created in Chapter 9. Again, half of a standard European ("pre-perfed") circuit board is enough to build this circuit.

## The Circuit And Its Function

It's easy to understand how the circuit works. By using the corresponding program from the companion CD-ROM, you can switch the individual light-emitting diodes on or off by a simple mouse click. The actions you perform are automatically transferred to the circuit so it always displays the setting you create in the program.

By enabling individual light-emitting diodes you can freely define characters, symbols, letters, etc. and output them on the matrix.

## Building The Circuit

The add-on circuit consists of the light-emitting diode matrix and two IC chips, which convert signals coming from the program so that the LEDs you have specified will light up. It should be obvious that a type of *decoding* must occur since the parallel port has only eight data lines, which cannot directly control the 35 light-emitting diodes.

The following layout shows the arrangement of the components.

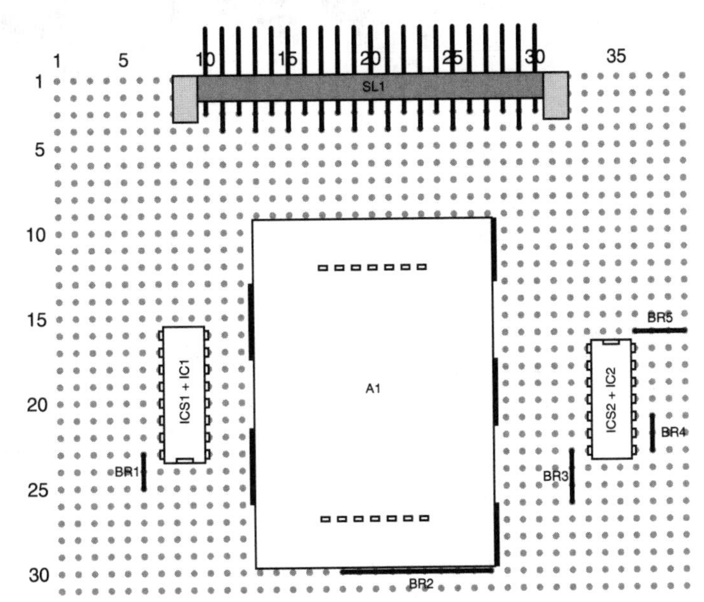


*The arrangement of the components*

When building the circuit, pay attention to the marking on the IC chips so that you don't solder them in wrong or insert them into the IC socket the wrong way.

Notice on the drawing that the conducting path bridge BR2 must be soldered very closely to the LED chip. Before you solder the LED matrix or the bridge, plug the two components into the board. Don't solder them to the board until both components have sufficient room. Otherwise, you might have a hard time plugging the bridge into the holes of the board if the matrix is covering them.

## Parts list

Refer to the parts list in the following table to determine which components you'll need for the programmable LED matrix

| Parts List | | |
|---|---|---|
| Label | Name | Component type |
| BR1 - BR5 | Conducting path bridges | Plastic-coated copper wire |
| IC1 | SN7442 | BCD for decimal decoder |
| IC2 | SN7404 | Inverter |
| ICS1 | IC socket 16 pin | IC socket |
| ICS2 | IC socket 14 pin | IC socket |
| A1 | TA20-11HWA 5x7-Light-emitting diode matrix (common anode) | Light-emitting diode matrix |
| SL1 | Pin strip 21pin | Pin connection |

The ICs are conventional logic chips that can be had for a reasonable price. The light-emitting diode matrix is a model with a *common anode*. Make sure you buy the right model, otherwise the individual light-emitting diodes (LEDs) won't light up, since they will receive the wrong signals.

## The Foil Side Of The Board

Refer to the following diagram for information on the conducting paths. Several conducting paths must be installed before the circuit will work.

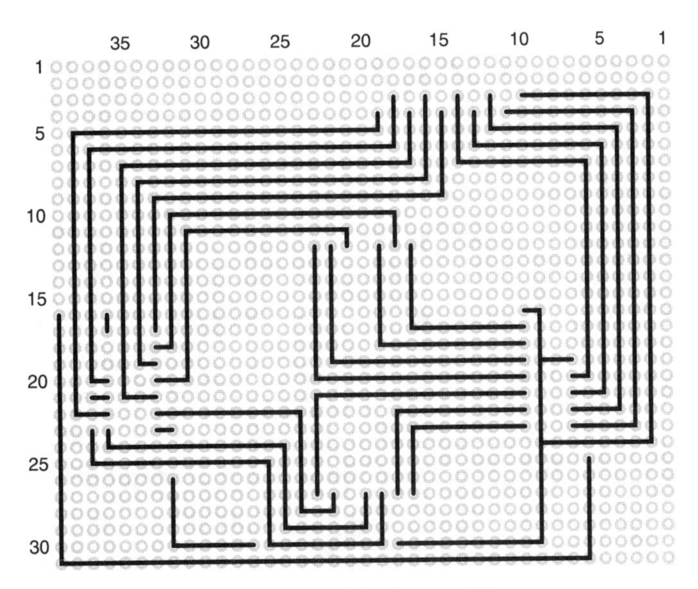


*The conducting paths of the LED matrix*

Once you have laid the conducting paths, the add-on circuit is ready for the basic circuit.

The finished circuit should look like the following illustration:

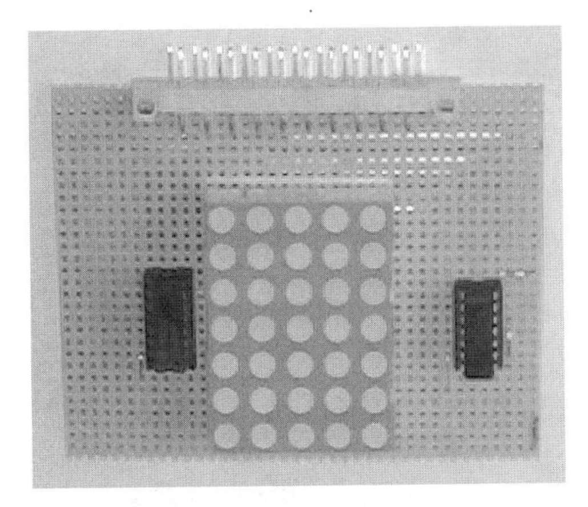

*The ready-to-use LED matrix*

## Using The Circuit With The Companion CD-ROM

Remember, we said at the beginning of this chapter that the program for this circuit has an important responsibility: It controls the light-emitting diodes of the matrix.

Use the program on the companion CD-ROM. As usual, you will find the Visual Basic source files in the directory.

After starting the program, the following screen appears:

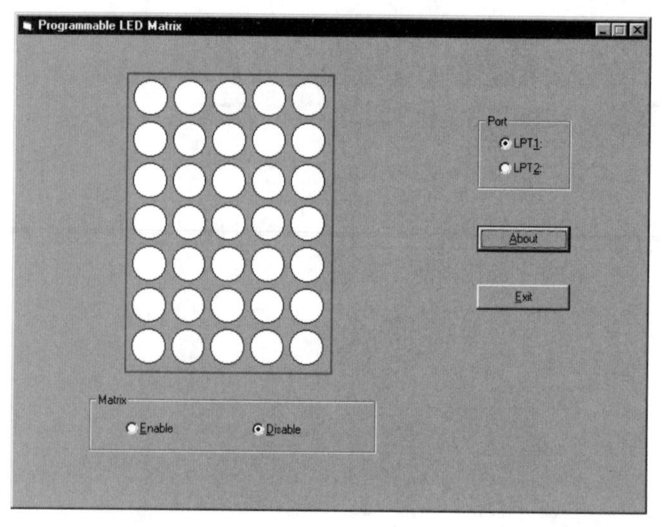


*You will use this program to control the LED matrix*

Along with the familiar elements for specifying which port you are using, the program information and exiting the program, you will find two additional items.

The top area represents the LED matrix, as you will find it in the circuit. The individual circles reproduce the LEDs, which can be switched on or off with a simple click of the mouse. You can do this at any time, whether the matrix is enabled or disabled.

Switch the LED matrix on and off through the two option buttons in the *Matrix* area. If the matrix is disabled, you can switch the LEDs on/off as desired. Your changes won't be displayed until you click on *Enable*. Other than that, all on/off cycles are passed on to the circuit directly.

## Using The Circuit

The following steps give you instructions on operating the circuit.

1. Plug the LED matrix add-on circuit with the pin strip into the socket strip of the basic circuit.
2. Connect the basic circuit to the parallel port of the computer using the printer cable.
3. Connect the external power supply to the basic circuit. Observe the nine (9) volt setting of the power supply and the set polarity. Then plug the power supply into the socket.
4. Run the program that goes with the circuit.
5. Enable the option for the port you are using—LPT1 or LPT2.
6. Click on the desired light-emitting diodes within the displayed matrix. If you accidentally clicked on the wrong LED, simply click on it a second time to switch it off.
7. Click on the *Enable* option in the *Matrix* area to have the light-emitting diodes on the circuit light up.

# Chapter 18: The Lotto Number Generator

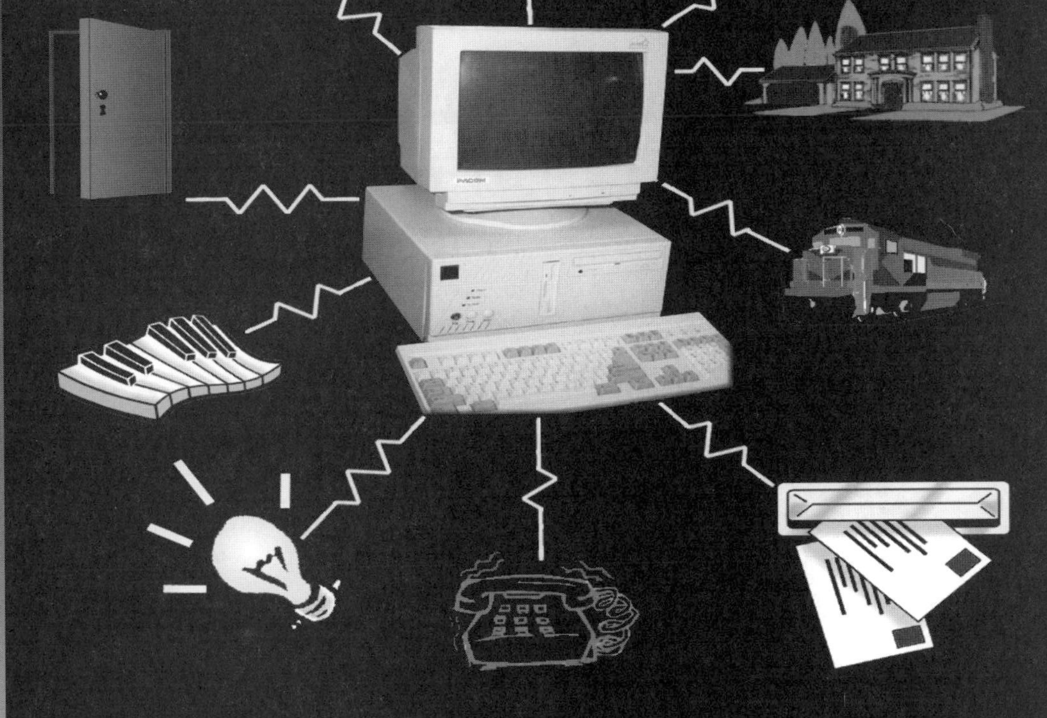

# Chapter 18: The Lotto Number Generator

Do you dream of winning the "big one" in a lottery? Regardless of whether you're a regular or a first time lottery player, you may have trouble deciding which numbers to pick for the drawing.

The circuit in this chapter is designed to help you select the (winning, we hope) numbers. It uses the basic circuit from Chapter 9 and a simple program you'll find on the companion CD-ROM, to generate numbers for you. You can then either use or ignore these numbers. In any case, please remember that we are not guaranteeing that you will win the lotto with these numbers.

Since this circuit was designed as an add-on circuit, half of a standard European ("pre-perfed") circuit board is enough to build it.

## The Function Of The Circuit

This circuit's task is to show the numbers generated by the program on a two-figure, seven-segment display. You can look at the numbers and decide whether to use them or ignore them.

**Important Tip**

You can also use the display for other purposes. To do this, you'll need to customize the program to suit your needs. For more information about customizing the program, see the "Using The Circuit With The Companion CD-ROM" section in this chapter.

## Building The Circuit

Two decoder chips are needed to display the numbers along with the two numeric displays. These chips ensure that the data furnished by the computer are converted into the correct signals for the display chips.

Since the data coming into the decoder chips is transferred by the basic circuit, make certain to have the basic circuit ready.

The components for this circuit should be assembled in the following positions.

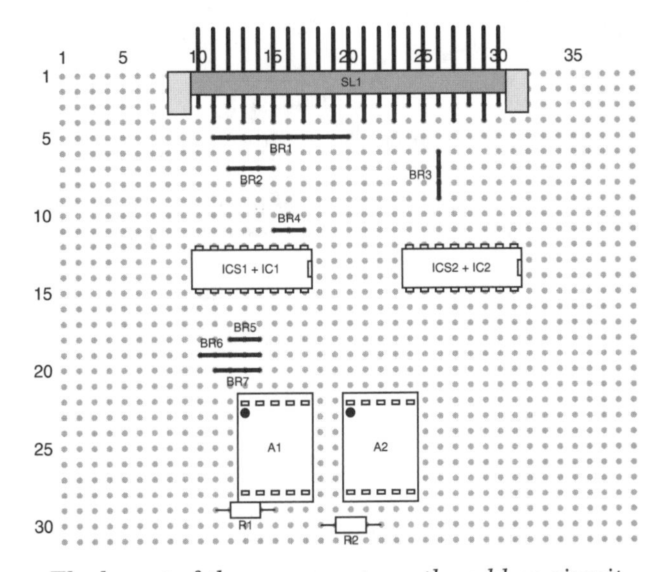


*The layout of the components on the add-on circuit*

When you solder the components to the board, be careful not to solder the seven segment displays incorrectly. Use the displays' decimal points to orient them properly.

Also, pay attention to the notch of the decoder IC and the IC socket, as with the other circuits.

### Parts list

The following table lists the parts list of the lotto number generator:

| Parts List | | |
|---|---|---|
| Label | Name | Component type |
| BR1 - BR7 | Conducting path bridges | Plastic-coated copper wire |
| R1 - R2 | 220Ω | Resistor |
| IC1, IC2 | SN 7448 | BCD for 7-segment decoder/display driver |
| ICS1, ICS2 | IC socket 16 pin | IC socket |
| A1, A2 | SC52-11GWA | 7-segment display (common cathode) |
| SL1 | Pin strip 21 pin | Pin connection |
| SL1 | Pin strip 21pin | Pin connection |

As you can see in the parts list, the circuit doesn't require many components. The basic circuit again has most of the necessary functions for this circuit.

The seven-segment display in the parts list is a specific part from an electronics firm. However, models from other manufacturers are also available so if you choose one of these other brands, make sure it has the right connections. It must have a common cathode. Also the pins of the chips must match those in the component drawing.

## The Foil Side Of The Board

Since the display chips must be connected to the IC connectors and the necessary data lines must be connected to the IC inputs, this circuit has a number of conducting path connections to be soldered.

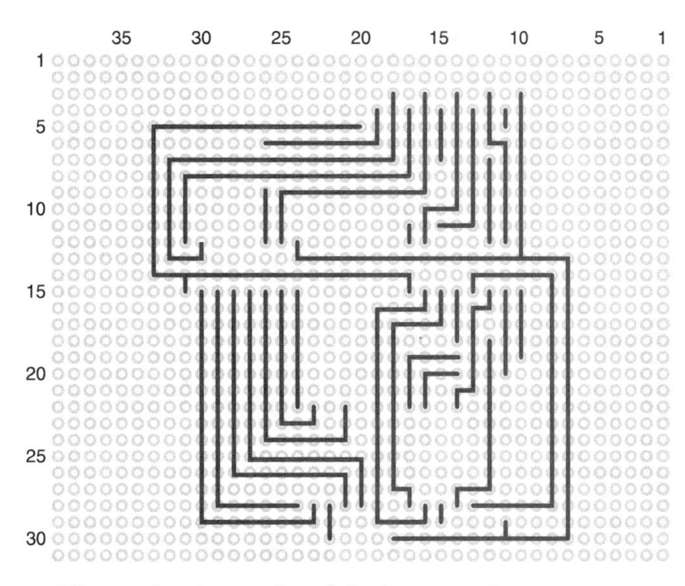


*The conducting paths of the lotto number generator*

The finished lotto number generator add-on circuit looks like this:

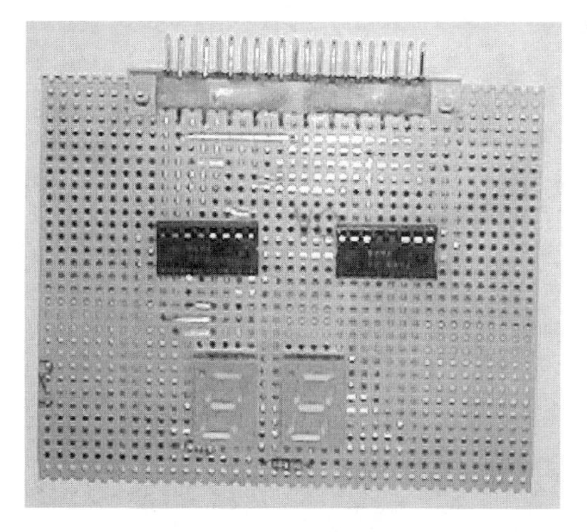

*The assembled lotto number generator circuit*

# Using The Circuit With The Companion CD-ROM

This circuit depends a great deal on the corresponding program from the companion CD-ROM. The program calculates the lotto numbers and transmits them to the circuit through the parallel port of your computer.

Install the program from the *chap_18* directory of the companion CD-ROM. After installation you can have the program display the lotto numbers.

As we mentioned, you can also use the circuit for other purposes. To do this, you will have to modify the program. If you wish to make changes to the program, use the source files from the *chap_18/source* directory as your basis.

Pay attention to the manner of data preparation and delivery. For each seven-segment display or for every decoder IC, a four-bit BCD value is transmitted to the port simultaneously. The resulting eight bits can then be sent to the decoder using parallel transmission.

When you start the installed program, you will see the following screen:

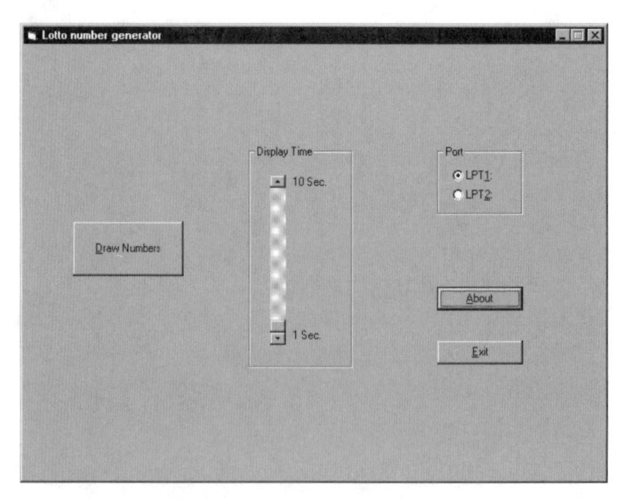


*The program for the lotto number generator*

The program doesn't have many options. You can begin number selection by pressing the *Draw Numbers* button. After that the display shows the generated numbers.

Use the scroll bar to adjust the time interval in which the numbers are displayed one after the other. You can choose settings ranging from one to ten seconds.

The six "normal" lotto numbers will be displayed sequentially. Before the additional number appears, the display is briefly disabled. That's how you can tell that the additional number is going to appear. Once the additional number has been displayed, the drawing is finished. The additional number continues displaying until a new drawing takes place.

The circuit doesn't require any other functions. The only other option you have is choosing between LPT1 and LPT2.

## Using The Circuit

The following steps give you instructions on operating the circuit.

1. Plug the add-on circuit for the lotto number generator into the socket strip of the basic circuit.
2. Use the printer cable to connect the basic circuit to the parallel port of your computer.
3. To feed the basic circuit with the external power supply, set the power supply to 9 volts with the correct polarity and then plug the power supply into the socket.
4. Start the installed program for the circuit.
5. Enable the correct option for the port you are using, either LPT1 or LPT2.
6. Start the "drawing" by clicking the *Draw Numbers* button. Adjust the speed of the display using the slider.

As you can see, this circuit provides a simple display of numerical values. You can use the circuit for different purposes. However, this would require making changes to the program using the source files.

If the display is too small for you, you can replace the seven-segment displays with larger display chips. You can get these from your electronics dealer.

# Chapter 19:

# The Electronic Clap Switch

# Chapter 19: The Electronic Clap Switch

This circuit lets you operate a switch by clapping your hands (we've tried hard not to add "Clap on! Clap off!" here). The circuit registers the noise of the clapping and transmits it to the computer, which activates a relay on the circuit. By connecting an electrical appliance to the relay you can switch it on or off by clapping your hands once.

The circuit transmits the data through the parallel port of the computer and requires an additional power supply. Although the circuit is larger than some others, you can still build it on half of a standard European circuit board.

## The Function Of The Switch

One component of the circuit is a small *loudspeaker* that we'll use as a *microphone*. Using this microphone, the circuit registers existing background noise near the component. The microphone converts the noises into electric impulses, which the circuit analyzes.

If a special impulse occurs, resulting from a clap, the circuit recognizes this and sends corresponding information to the computer. The software running on the computer reacts to the clap and switches a relay on the board.

With the help of the closing or opening contacts of the relay you can switch an electrical appliance on or off.

To operate the circuit, you will need the external power supply and the flat ribbon cable with the Centronics plug that you also used for the basic circuit.

# Building The Circuit

The circuit has over thirty components that you can easily solder to the board. Also you must attach the small loudspeaker we mentioned earlier. We'll tell you more about that later.

First, let's build the circuit. The following drawing shows the layout of the components.

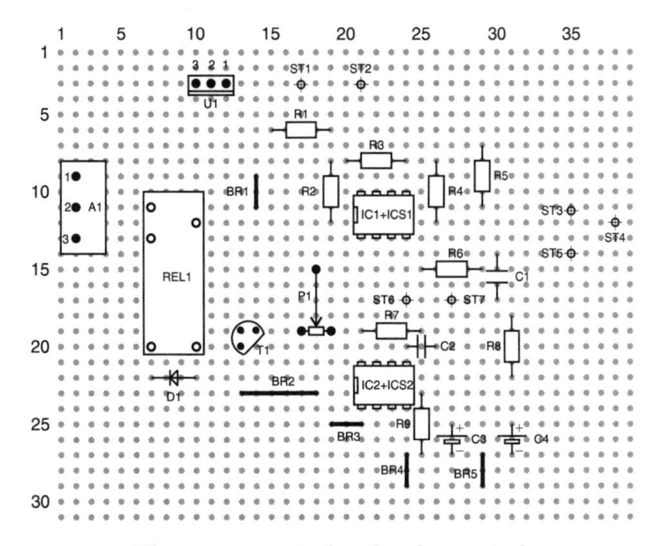


*The components for the clap switch*

## Parts list

Refer to the parts list in the following table to determine which components are used in this chapter.

| Parts List | | |
|---|---|---|
| Label | Name | Component type |
| BR1 - BR5 | Conducting path bridge | Plastic-coated copper wire |
| R1 | 2.2 kΩ | Resistor |
| R2 | 4.7 kΩ | Resistor |
| R3 | 470 kΩ | Resistor |
| R4 - R6 | 100 kΩ | Resistor |
| R7 | 47 kΩ | Resistor |
| R8 | 220Ω | Resistor |
| R9 | 1 kΩ | Resistor |
| P1 | 22 kΩ | Potentiometer (horizontal) |
| C1 | 22 nF | Condenser |
| C2 | 0.1 µF | Condenser (Elko) |
| C3, C4 | 2.2 µF | Condenser |
| T1 | BC546B | Transistor |
| D1 | 1N4007 | Diode |
| IC1 | NE555 | Timer IC |
| IC2 | NE532 | Operational amplifier IC |
| ICS1, ICS2 | 8 pin IC socket | IC socket |
| U1 | 7805 | Voltage regulator |
| REL1 | 5-V Power relay 1 x UM | Card relay |
| A1 | 3 pin connecting terminal | Connecting terminal |
| ST1 - ST7 | Edge connector (female) | Pin connection |

## Building the circuit

When soldering the components to the board, mind the following items—otherwise the circuit may not work properly:

1. Pay attention to polarity when working with the diode.
2. The two ICs must be correctly inserted into the socket, and the IC socket must be properly soldered (as shown in the drawing).

3. Since the relay in this circuit must be used exactly like the one in Chapter 15, refer to that chapter for your relay data, and buy the right one.

4. You are also familiar with the connecting terminals from Chapter 15. Read important information about this component in that chapter.

5. The potentiometer used in this circuit is not a spindle potentiometer. Instead, it is a simple, more reasonably priced model. When purchasing it, be sure to get a horizontal model, so that you can use the component as shown in the drawing. Make sure the dimensions of the potentiometer match the dimensions shown, since potentiometers come in different sizes.

6. When soldering condensers C3 and C4, pay attention to the polarity of these electrolytic condensers.

After soldering the components, you also need to solder the necessary conducting paths on the rear side of the board.

## The Foil Side Of The Board

Refer to the following diagram for information on the conducting paths.

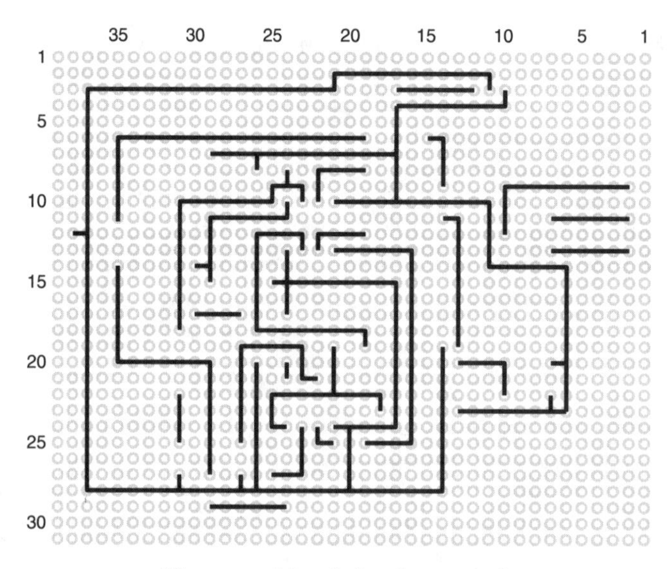


*The rear side of the clap switch*

There is a little work left after you finish soldering the conducting paths.

## Additional Work On The Circuit

Additional elements must be connected to the board through the female edge connectors. Although you're familiar with some of the work, there are also some new tasks you'll need to do. Fortunately, they're similar to what you have already learned.

First, let's turn our attention to the loudspeaker we mentioned at the beginning of the chapter.

This is a reasonably priced small loudspeaker with an internal resistance of 8 ohm. You can find such a loudspeaker at any electronics shop.

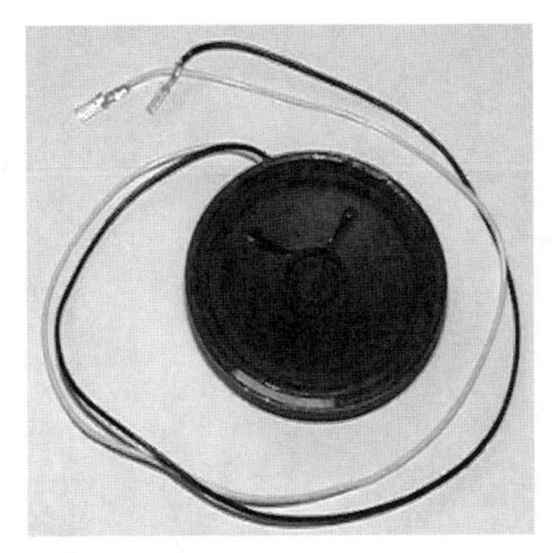

*A small loudspeaker serves as a microphone*

Follow these steps to connect the loudspeaker to the circuit.

1. The loudspeaker has two contacts. You need to solder a two-conductor cable to these contacts so that you can position the loudspeaker more or less independently from the circuit.

   Solder two male edge connector to the conductors at the other end of the cable. These plug into the female edge connectors, ST6 and ST7. Polarity is unimportant.

2. The external power supply is attached to the two female edge connectors, ST1 and ST2. Unlike the loudspeaker, polarity is important here. Plug the positive wire into female edge connector ST1, plug the ground wire into female edge connector ST2.

3. Since the circuit exchanges its information through the parallel port of the computer, the flat ribbon cable with the Centronics plug must be plugged into the three free female edge connectors. Observe the following arrangement:

| | |
|---|---|
| ST3 | Wire 3 |
| ST4 | Wire 2 |
| ST5 | Wire 1 |

Once you have completed these tasks, the circuit is ready to be used.

*The finished circuit of the clap switch*

## Using The Circuit With The Companion CD-ROM

This circuit also requires software. You'll find the installation files on the companion CD-ROM in the *Chap_19* directory. You can get the program's Visual Basic files from the subdirectory if you want to change the circuit's function. For example, you could modify the source files to start a specific program when someone claps.

First, though, let's turn our attention to the existing program. Once you have installed it, you can start it up. The following figure shows the program with all the possible functions for this circuit.

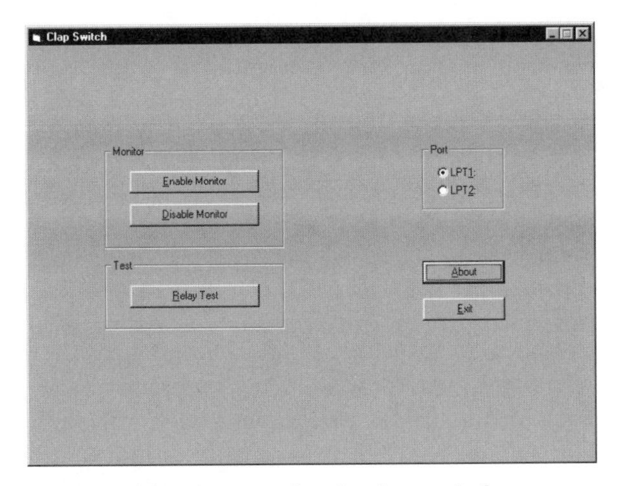


*The program for the clap switch*

You are already familiar with selecting which port you are using (LPT1 or LPT2).

The circuit lets you test the relay by clicking the *Relay Test* button. As soon as you click this button, the circuit switches on the relay and then switches it off after about 2 seconds.

To switch on the circuit, click the *Enable Monitor* button. Only then will the program react to the impulses of the circuit.

To switch the circuit off again, click the other button, *Disable Monitor*.

The program has no influence over the sensitivity of the circuit. The potentiometer, P1, is on the circuit for that purpose. Depending on which direction you turn the potentiometer, the circuit will become more or less sensitive.

## Using The Circuit

The following steps give you instructions on operating the circuit.

1. First, connect the cable of the loudspeaker (microphone) to the female edge connectors on the circuit.

2. Connect the circuit to the parallel port of the computer using a printer cable.

3. Plug the two wires for the external power supply into the female edge connectors of the circuit. Watch the polarity. Plug the power supply into the socket.

4. Start the program and select the correct port. Click the *Relay Test* button to test the relay.

5. Now clap near the loudspeaker. If the circuit doesn't react, change the position of the potentiometer with a small screwdriver. This is how you set the sensitivity of the circuit. Experiment with different settings. Set the potentiometer to have the circuit react only to the sound of clapping.

After finding the right setting, the circuit is ready to use. Connect the wires of the item to be controlled to the contacts of the relay or the connecting terminals.

Between terminals 1 and 2 is an normally closed, while a normally open is between terminals 1 and 3.

Before connecting the appliance to be controlled to the clap switch, read the following warnings.

## Be careful with this circuit

Since the circuit is designed to switch lamps and similar appliances on and off, you must observe certain safety measures. Your negligence could result in injuries.

Read the safety notes at the end of Chapter 15. Everything discussed there also applies for this circuit. Be sure to observe the precautions discussed there as well as this warning.

**Important Tip**

If you intend to control high voltage through the terminals, find a nonconducting case and attach it around the circuit. This will protect you from direct contact with the conducting paths. You should also consider this when controlling alternating voltage, even with lower voltage values.

# Chapter 20:

# The Flow Indicator

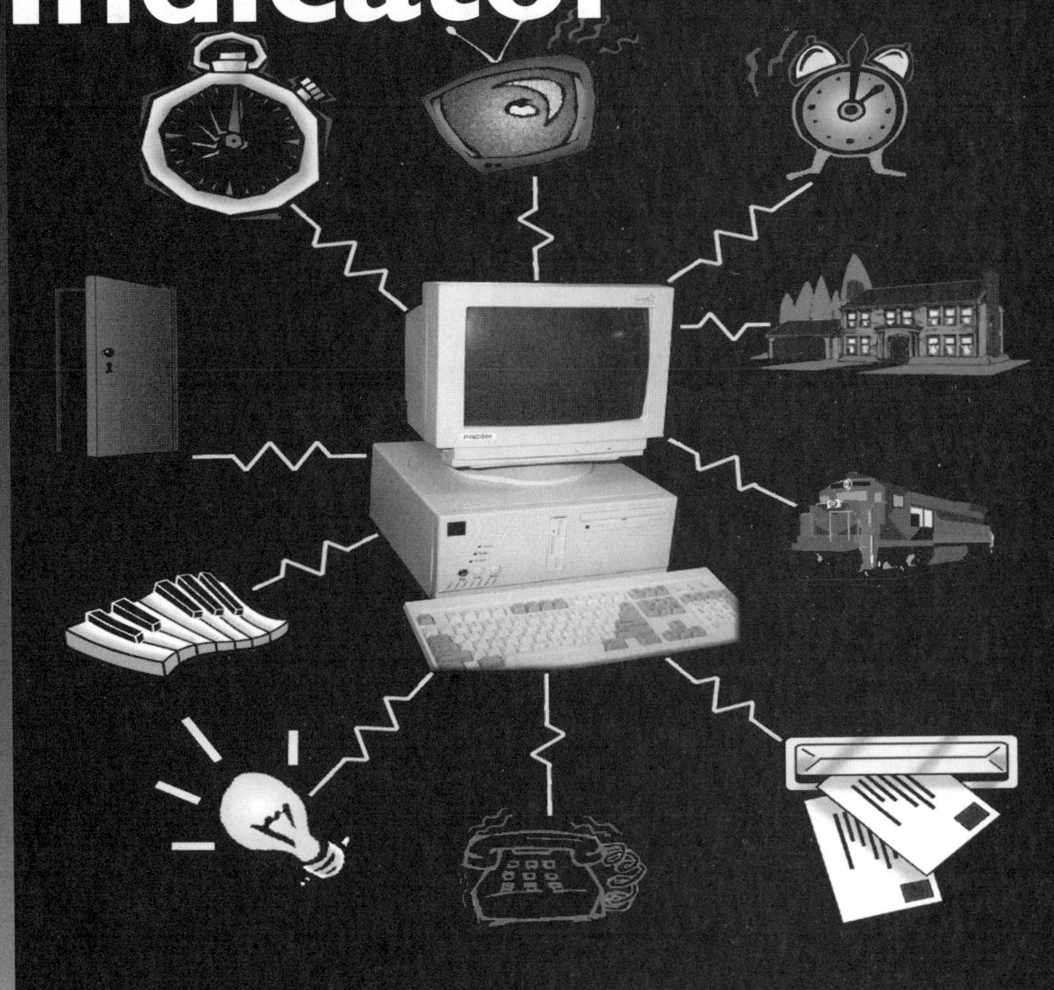

# Chapter 20: The Flow Indicator

The main item of the flow indicator circuit is a *photoelectric beam*. This circuit registers when the beam is interrupted and passes this information to the software through the parallel port. The software uses a counter to log the number of photoelectric beam interruptions. For example, you could use this circuit to count the number of laps of a model train or a car on a race track.

The circuit requires at least one more power supply. You also have the option of running the circuit with a second power supply (a battery). We'll tell you more about that later. Since the circuit consists of two parts (transmitter circuit and a receiver circuit), you must prepare half a standard European circuit board for the receiver and one quarter of a standard European circuit board for the transmitter.

## The Circuit And Its Function

As soon as it has power, the transmitter circuit sends a light (which you cannot see) to the receiver. The receiver circuit picks up this light and produces a corresponding circuit state, which is transmitted to the software on the computer.

If the light beam from the transmitter to the receiver is interrupted, e.g., by a passing object, the receiver changes its circuit state. The software registers this, analyzes it and gives a visual indication of this change. When the interruption of the light beam is no longer there, the original state of the circuit is restored.

However, the circuit can only function when the transmitter and the receiver are positioned so that the transmitter diode can send directly to the receiver, and the receiver is able to receive the light. So the two circuits have to be arranged facing each other.

In some situations discharge from the power supply (like static electricity) can interfere with the receiver. Using a second power source, a battery, can avoid this.

This also means you won't have to run wires between the transmitter and receiver. So for example, if you want to monitor traffic through a doorway, you won't have to run wires across the doorframe.

The receiver part of the circuit will be fed by the external power supply with which you are familiar from previous chapters. You will also need the flat ribbon cable with the Centronics plug for the parallel port.

## Building The Circuit

Since you are building both a transmitter circuit and a receiver circuit, we've divided construction of the circuit into two sections.

### The transmitter is first

First, let's turn our attention to the transmitter, which will be built on a quarter of a standard European circuit board. The following drawing shows the few components you will need for the transmitter.

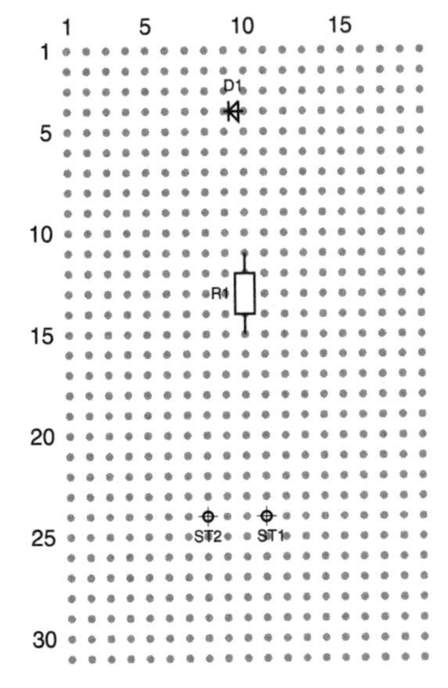


*The component layout for the transmitter*

## Parts list for the transmitter

Refer to the parts list in the following table to determine which components are used in building the transmitter.

| Parts List | | |
|---|---|---|
| Label | Name | Component type |
| R1 | 56Ω | Resistor |
| T1 | Infrared diode SFH 409 | Diode |
| ST1, ST2 | Edge connectors (female) | Pin connection |

## Building the transmitter

Although building the transmitter is easy, remember the following:

1. The transmitter diode looks like the normal light-emitting diodes you are familiar with from other circuits. The polarity of this diode is the same as those others. However, don't push the diode so far through the board that it lies flat on it. You will need to bend the diode by 90° so that the transmitter beam can be aimed directly toward the receiver part without having to stand the board upright.

2. The female edge connectors ST1 (positive) and ST2 (negative) link the power supply to the circuit. You can either attach a cable that taps power from the receiver board, or you can use a 9-volt battery. There is a special connection cable for this type of battery, to which you can solder male edge connectors. Using a voltage regulator, you can limit the voltage to 5 volts. You can also place a switch in the lead, so that you can switch the transmitter on and off with the switch. This way you can feed the circuit with power from the battery. However, pay attention to polarity.

3. When mounting a switch, make sure you record which positions of the switch turn the circuit on and off. That way you will always know when the circuit is transmitting and when it isn't.

## The foil side of the transmitter

Connect the components of the transmitter module as shown in the following drawing.

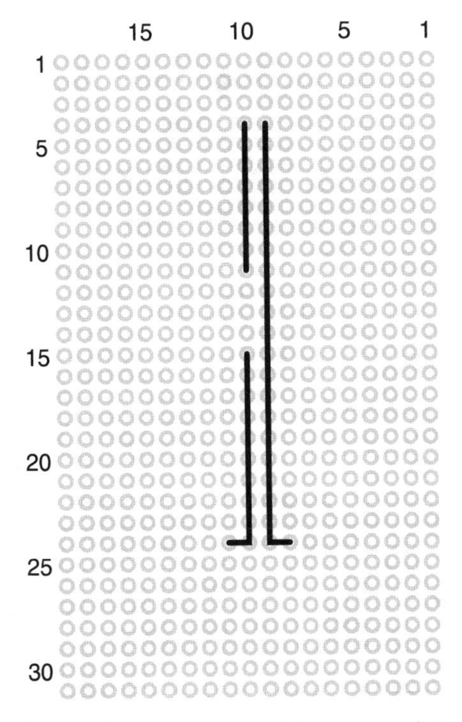


*The conducting paths of the transmitter*

After you finish building the transmitter, the circuit will look like the following illustration:

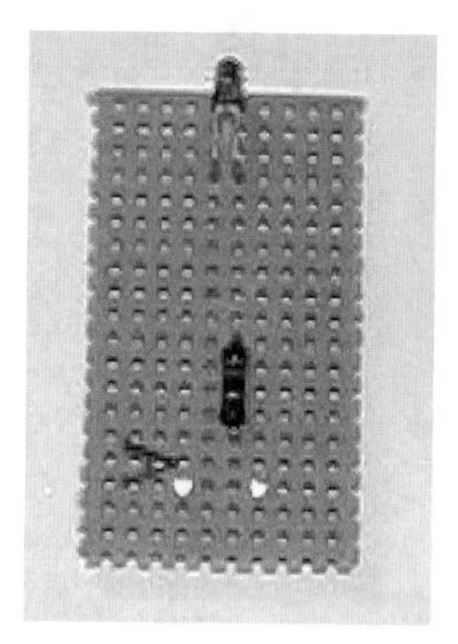

*The finished transmitter*

## The receiver is next

When you've finished the transmitter, start on the receiver module. This part of the circuit doesn't have many components either so use half a standard European circuit board. The following drawing shows where to place the components for the receiver.

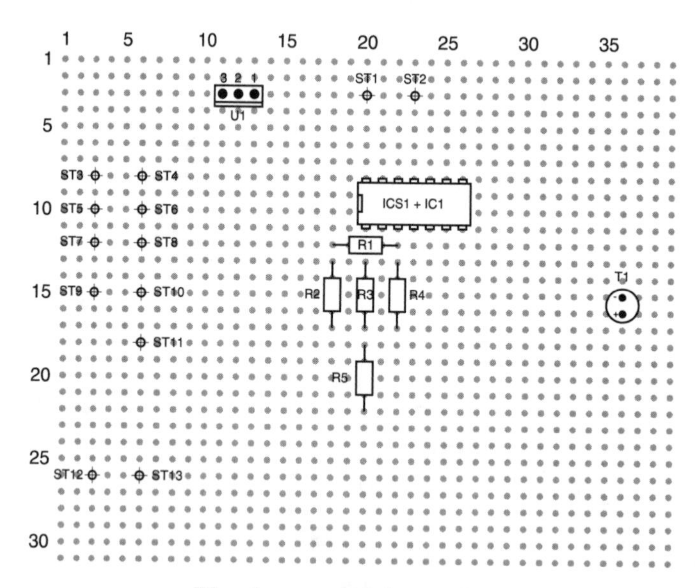


*The elements of the receiver*

## Parts list for the receiver

Because there are so few components, building this part of the circuit is easy. The following table lists the parts you'll need:

| Parts List | | |
|---|---|---|
| Label | Name | Component type |
| R1, R4, R5 | 10 kΩ | Resistor |
| R2, R3 | 100 kΩ | Resistor |
| IC1 | LM324 | 4x Operational amplifier |
| T1 | SFH 309 | Phototransistor |
| U1 | 7805 | Voltage regulator |
| ST1 - ST13 | Edge connectors (female) | Pin connection |

When building the circuit, make sure that you bend the receiver transistor by 90°, as you did with the transmitter diode. Both components, the transmitter diode and the receiver transistor, should be bent at the same level.

When installing the phototransistor, pay careful attention to the polarity. The plus and minus signs in the drawing indicate the correct orientation. To help you install the component properly, we refer to a positive and a negative pole. Devote your undivided attention to the following explanation about the polarity of the phototransistor.

Though the transistor looks similar to a diode, the polarity for a phototransistor is exactly the opposite of the polarity for a light-emitting diode. The longer terminal wire is the cathode—the negative pole. The short terminal wire on the phototransistor is the anode, or the positive pole. Keep this in mind when you install the phototransistor.

If you install the phototransistor incorrectly, the receiver module will only respond to the transmitter if it is at close range (a distance of only 1 to 5 mm). If you notice this phenomenon, check whether you installed the component correctly.

## The foil side of the receiver

Here are the conducting path connections for the rear side of the receiver board:

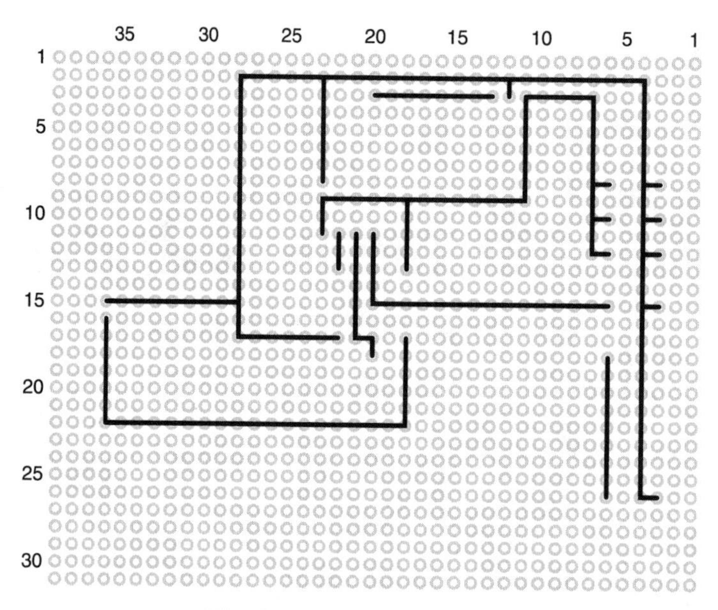


*The foil side of the receiver*

After building the board, you can connect the necessary cables.

1. Plug the two wires of the external power supply into female edge connectors ST1 (positive) and ST2 (negative).

2. The two female edge connectors ST3 (positive) and ST4 (negative) are designed to supply the transmitter with power. If you don't use a battery, then connect a two-conductor cable from these two female edge connectors to the two appropriate edge connectors of the transmitter module. Be sure to observe the polarity of the female edge connectors and the two wires. Also, remember that the transmitter, if you attached a switch, will have to be switched on in order to transmit.

3. Plug the flat ribbon cable for the parallel port into female edge connectors ST9 and ST10. Notice the arrangement in the table to the right.

| ST9 | Wire 2 |
|---|---|
| ST10 | Wire 19 |

We'll ignore the other female edge connectors in the circuit. They have been integrated in anticipation of the next chapter's circuit.

After you finish these tasks, the receiver and the entire circuit are ready for use.

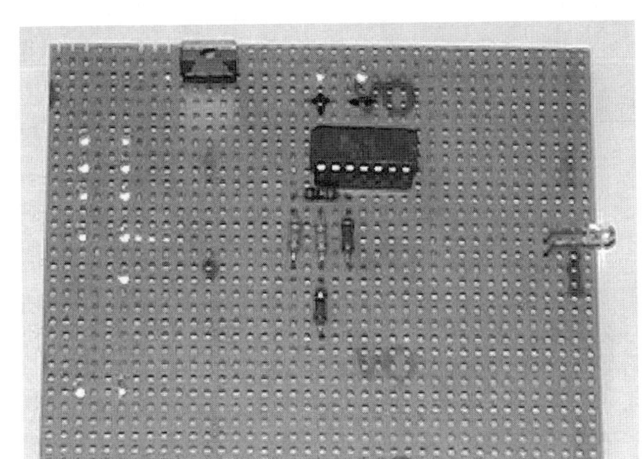

*The receiver module of the flow indicator*

## Using The Circuit With The Companion CD-ROM

The circuit's software program monitors and analyzes the signals of the flow indicator. The software is located on the companion CD-ROM in the *Chap_20* directory. Use the files in this directory to install the software. If you are looking for the source code files, you will find them in a subdirectory called *Source*.

After you finish installation and start the program, the following screen appears.

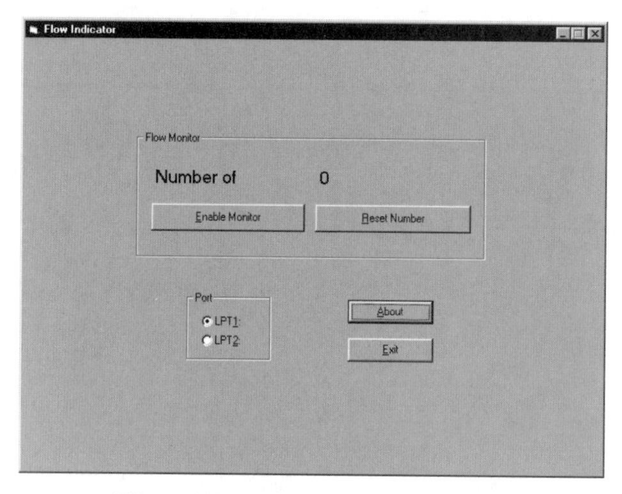


*The software for the flow indicator*

The option for setting the port is mandatory. Select the port you are using, either LPT1 or LPT2.

In the *Flow Monitor* area, a number indicates how often the light beam is interrupted. The two buttons there, *Enable Monitor* and *Reset Number,* allow you to enable, disable or reset the display.

## Using The Circuit

This circuit requires more setting up than you are accustomed to from the previous circuits. Not only do you have to create the required cable connections, you also have to ensure that the receiver captures the light of the transmitter module, which can be difficult because you cannot see the light. Nevertheless, here are the steps:

1. First, connect the receiver module to the computer with the printer cable.
2. Provide the recover module with the external power supply. Keep the polarity in mind. Plug the power supply directly into the socket.
3. Start the software and set the correct port option.

4. Set up the transmitter module about 4 inches from the receiver. Make sure that the diode can beam rays to the receiver and that the receiver transistor is directed toward the diode.

5. Supply power to the transmitter either by connecting it to the receiver board or to a battery.

6. Close the switch to turn on the power and allow the DIODE to begin transmitting. If you didn't install a switch, the diode begins transmitting as soon as the power supply is connected.

7. You can test whether the circuit is functioning properly by clicking the *Enable Monitor* button and holding your hand between the transmitter and the receiver. The counter will do its job.

8. Once you have determined that the circuit is functioning properly, you can begin placing the two modules where they will be used. After placing them, repeat these procedures. Keep the following general comments in mind.

   - You can use the circuit for various tasks: as a lap scorer of a model race track or for any application where moving parts can result in changes of state in the circuit.

   - Do not place the two modules too far from each other. To test for the maximum distance, operate the circuit as described above and then keep moving the transmitter further away.

   - If you're powering the transmitter with a battery, make sure this is switched off when the circuit is not in operation. Otherwise, the transmitter will continue transmitting until the battery is discharged. You'll be surprised when the circuit doesn't work for you the next time.

   - When using the two modules, the receiver reacts to the transmitted infrared light of the infrared diode. Since normal daylight or heat developing devices, such as a powerful electric heater, influence the infrared light of the diode, be sure to use the circuit only where such influences don't come into play.

If you discover that the receiver also reacts to normal light, there are two things you can do. For one, wrap the phototransistor with opaque material (material that is impervious to light), so that the light cannot directly affect the component. You can easily achieve this with a black tube. It doesn't matter if the tube projects beyond the transistor. The transmitter can then easily shine into the tube from the open end.

If disturbances continue to occur, cover the input area with a red filter screen. You may be familiar with this type of screen from your television set. The receiver area of the remote control is often furnished with such a screen to filter out the normal daylight. You can find this filter screen at an electronics shop.

# Chapter 21:

# The Speed Indicator

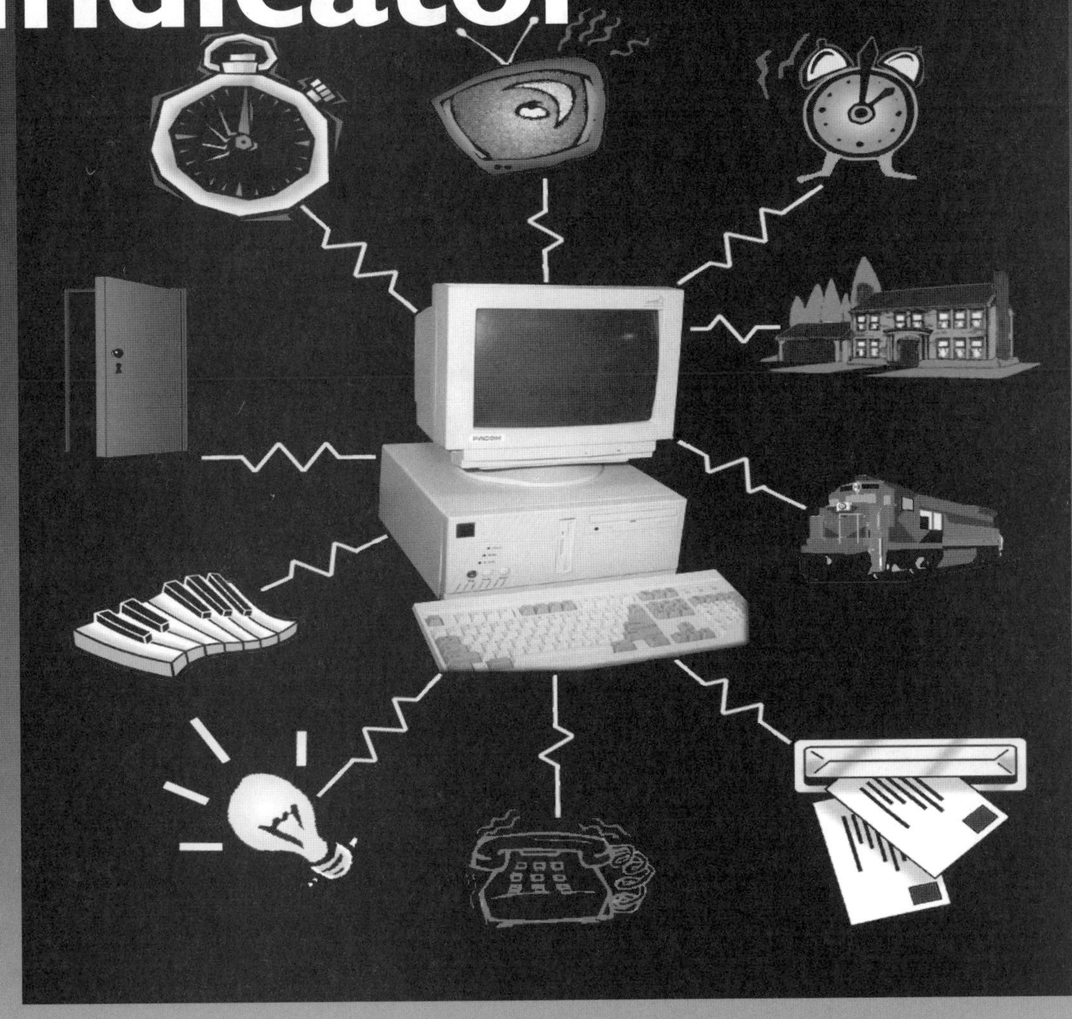

# Chapter 21: The Speed Indicator

The circuit you'll build here is based on the same principle as the circuit from Chapter 20. It uses the same photoelectric beam, even though this circuit is designed to measure speed. To do so, this circuit requires two photoelectric beams.

This circuit will also require an additional power supply. You can supply power to this circuit with either the power supply on the receiver board or batteries.

The circuit is made up of four boards, including the two circuits from Chapter 20. The most important board connects to the parallel port of the computer and contains a receiver module. An additional receiver module, built on a separate board, is connected by cable to this board.

Along with the two receiver boards, the circuit requires two identical transmitters. Since the transmitters are the same as the one you built in Chapter 20, you can use that transmitter and build one more.

This circuit is designed for use with toys such as model trains; it is not possible to use this circuit to measure speeds of real full size cars.

## The Circuit And Its Function

You know from Chapter 20 that the photoelectric beam consists of a transmitter and a receiver. If the light beam between these two modules is interrupted, an impulse is sent to the software.

In this circuit, two photoelectric beams are used so that the two light beams are interrupted one after the other by an object in motion, for example, a model train. This requires that the path (the tracks) of the object's (the model train's) motion runs through the two photoelectric beams.

This process generates two impulses from the photoelectric beams. Speed is calculated by the delay between the two impulses. However, the intended speed calculation requires that the distance between the two photoelectric beams (set as exactly as possible) be communicated to the software.

Using a model train (for a simplified description), the measuring process works like this: The photoelectric beams are placed at a defined distance from each other, e.g., 20 inches, on a level piece of track. When the train passes through the first photoelectric beam, the computer registers and stores this information. When the train passes through the second photoelectric beam, the computer receives another impulse. The software registers this as well and then calculates the time between the two impulses. Based on the distance between the beams and the time between impulses, the program calculates the speed, which it then displays to you.

The following section talks about the steps for building and operating the circuit .

## Building The Circuit

As you have learned, the circuit consists of four parts, which are built on the same number of boards. Let's turn our attention to the transmitters first.

### The transmitters are first

Since the transmitters for the photoelectric beams used in this chapter are identical to the transmitter that you built in Chapter 20, you can use that transmitter here as well.

Since you need two transmitters, build a second one as described in Chapter 20. The second transmitter is absolutely identical to the one from Chapter 20.

If you don't want to supply either of the two transmitter modules with power from batteries, the receiver board has female edge connectors for both modules to receive the necessary power.

Get all the necessary information for building the transmitter boards from Chapter 20, including the parts list and the component layout. The transmitter modules shouldn't give you any trouble.

## The receivers are next

The receiver modules consist of two different boards. One board is from Chapter 20. That board contains all the necessary components for this circuit. The second receiver is quite similar to the board from Chapter 20, particularly in the components used.

If you have already built the receiver from Chapter 20, you don't have to build it a second time. Otherwise, build it first. The instructions for building this receiver are in Chapter 20.

Once the board is finished, you can provided it with the necessary cable connections.

1. Connect the external power supply for the circuit to female edge connectors ST1 (positive) and ST2 (negative).
2. Female edge connectors ST3, ST5 and ST7 (all negative) and female edge connectors ST4, ST6 and ST8 (all positive) are available for supplying the transmitter modules or the second receiver module with power. It's up to you to choose which pair of female edge connectors you use for which module. You don't even have to connect all the boards this way. You could also build the circuit so that, for example, the second receiver board is supplied with power this way, while the two transmitters are powered by batteries.
3. The two female edge connectors ST12 and ST13 are where you connect the signal cable of the second receiver. The negative wire goes to ST12, while the positive wire goes to ST13.
4. The flat ribbon cable for the parallel port uses female edge connectors ST9 to ST11. Connect the following wires with the appropriate female edge connectors:

| | |
|---|---|
| ST9 | Wire 2 |
| ST10 | Wire 19 |
| ST11 | Wire 23 |

After you finish building and connecting the main board, turn your attention to the second receiver, which is the last to be built.

## The second receiver

The structure will be familiar to you, since it is not much different than the main receiver.

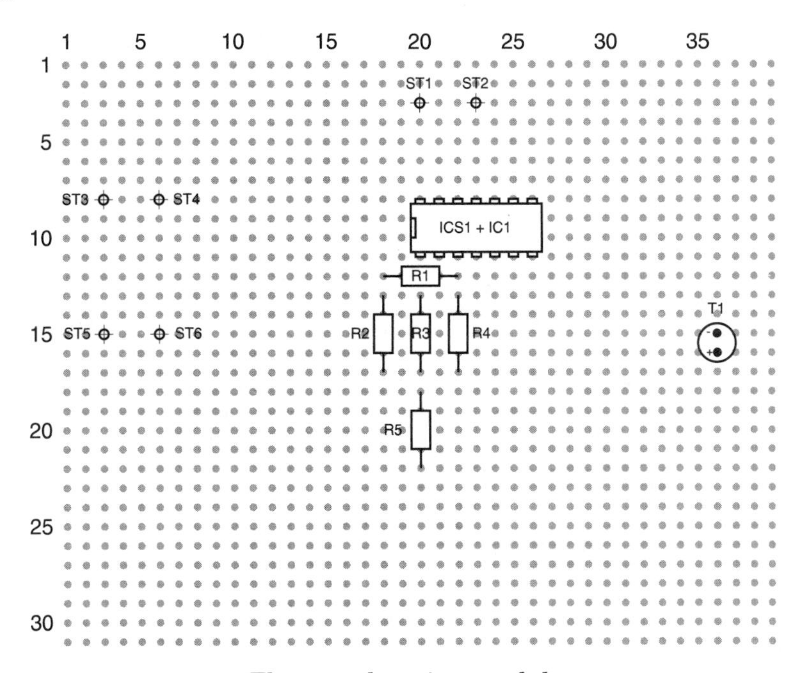


*The second receiver module*

## Parts list for the second receiver module

Since the same elements are used in the main receiver, the parts list for this module consists of familiar components.

| Parts List | | |
|---|---|---|
| Label | Name | Component type |
| R1, R4, R5 | 10 kΩ | Resistor |
| R2, R3 | 100 kΩ | Resistor |
| IC1 | LM324 | 4x Operational amplifier |
| T1 | SFH 309 | Phototransistor |
| ST1 - ST6 | Edge connectors (female) | Pin connection |

Connect the components to each other as shown in the following illustration:

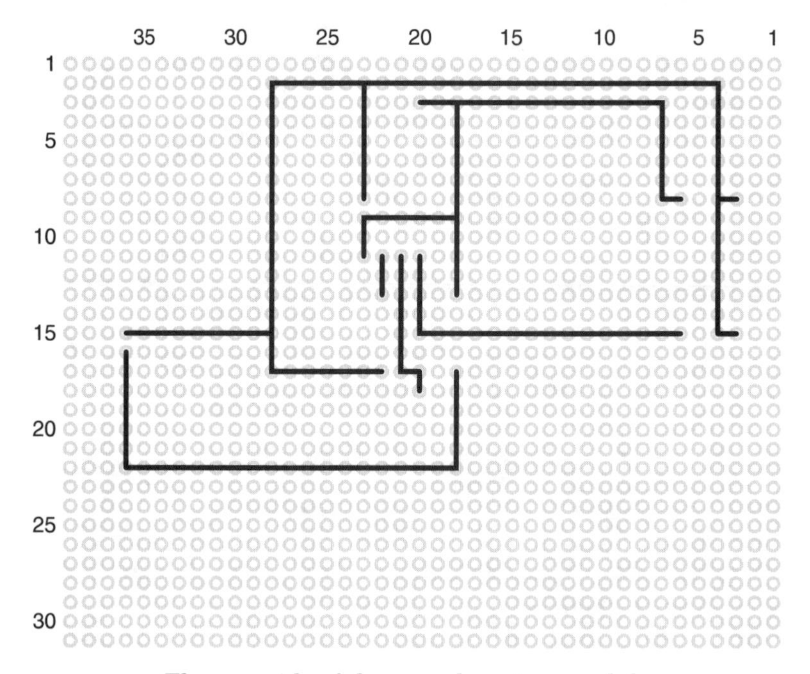


*The rear side of the second receiver module*

The board has six female edge connectors, used for the power supply and for the signal line.

1. Plug the external power supply into female edge connectors ST1 (positive) and ST2 (negative). It doesn't matter whether they are from the main receiver board or from a battery. However, note the polarity and remember you must use a voltage regulator to limit to 5 volts.

2. The second pair of female edge connectors, ST3/ST4, supplies 5 volts for a transmitter. You can tap 5 volts at ST4 positive and ST3 negative. If you don't feed the transmitter with power from here, you can ignore this pair of female edge connectors.

3. The ST5/ST6 pair of female edge connectors is in charge of signal transfer to the main receiver board. Here you have to plug a two-conductor wire to the main receiver board. Connect the line from female edge connector ST5 of the second receiver to female edge connector ST12 of the main receiver board, and connect the line from female edge connector ST6 of this module to female edge connector ST13 of the main module.

As soon as this board is built, you can begin using it.

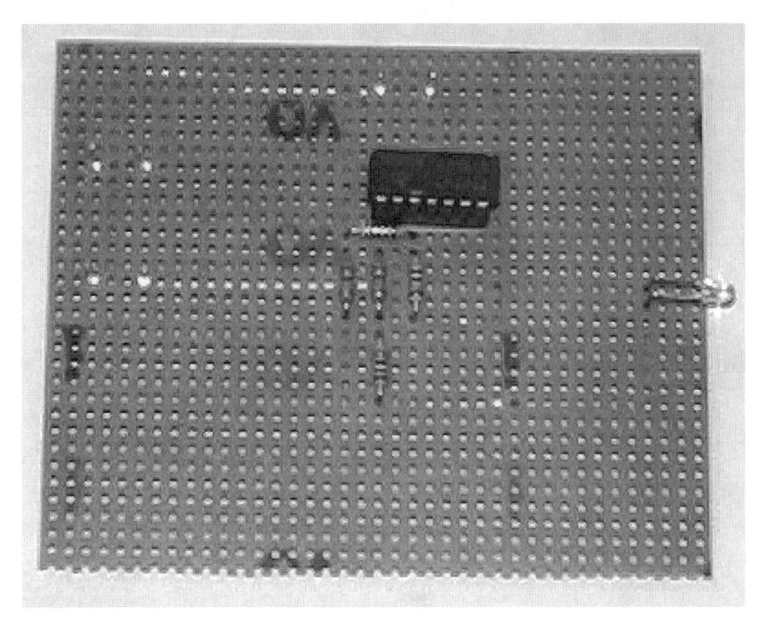

*The finished second receiver module*

## Using The Circuit With The Companion CD-ROM

The software for this circuit calculates the measured speed. You'll find it on the companion CD-ROM in the *Chap_21* directory. Once you have installed the program on your computer, you can put the circuit to use. You can start additional functions within the source code files that are also on the companion CD-ROM.

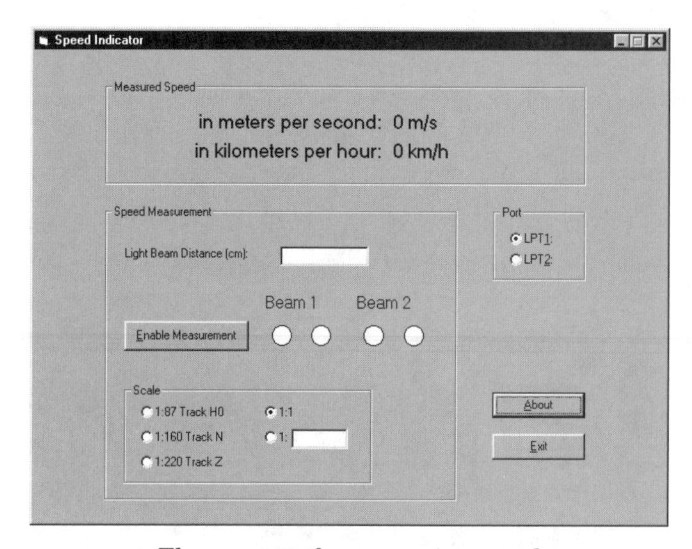


*The program for measuring speed*

As with all the circuits operated at the parallel port of the computer, this program has an option for specifying port LPT1 or LPT2. To enable the software to correctly calculate speed, you have to specify the distance between the two photoelectric beams. Enter the distance in the *Distance Between Photoelectric Beams* input line.

Click the *Enable Measurement* button to ready the program to perform a speed measurement. For example, before the train passes through the photoelectric beams, you have to click this button to prepare the measurement. As soon as the train has passed through the two photoelectric beams, the program analyzes the information it receives from the circuit and displays the result to you. Before you can perform another measurement, you have to click the *Enable Measurement* button again, to restore the readiness of the circuit and the program.

The program outputs the computed speed within the *Measured Speed* area. The program displays the speed in both meters per second (m/s) and kilometers per hour (km/h).

So you can compare the measured speed of the model to a real train, you can specify a scale factor. If you are measuring the speed of a 1:16 model, then according to the measured speed, the scale speed is also output on the screen.

You can freely define a factor in the *Scale* area, though some default model train scales are provided. The scale you choose is then automatically taken into account in the calculation. The *1:1* option lets you output the measured speed without a scaled correction.

You can check the reactions of the photoelectric beams by using the two lights under *Beam 1* and *Beam 2*. As soon you click on *Enable Measurement*, the green lights light up on both photoelectric beams. That means that the circuit is read to use. When the train passes through the first photoelectric beam, the corresponding light changes from green to red. The same thing happens when the train passes through the second photoelectric beam. From this you can easily identify the status of the circuit and software.

Before you enable measurement, the photoelectric beams have to be ready to use. That means that both receivers must receive the signal from the transmitters. If this is not the case, you will get an error message, and it is not possible to perform a measurement. If the second photoelectric beam is ready to use, but the first one is not, you will be able to tell, because the light of the first photoelectric beam immediately changes to red after you click the *Enable Measurement* button. In this case, the program won't determine the correct result either.

## Using The Circuit

A number of steps are necessary to operate the circuit for measuring speed. You are already familiar with some of them from Chapter 20. For this reason, we won't give you a detailed explanation of those steps a second time. If necessary, refer to the last chapter.

1. First you need to align the circuits. This involves, among other things, getting the receiver and transmitter boards in position. Set the two photoelectric beams up so that each transmitter sends to its receiver. Measure the distance between the two photoelectric beams and write it down.

2. Provide all the modules with power, as described earlier. Before plugging the external power supply into the socket, make sure that the required signal lines between the two receiver boards are set up and that the flat ribbon cable is connected with the parallel port. Once you have done this, you can switch on the external power supply as well as the individual boards. In particular, the transmitters have to be turned on via switches, if installed.

3. Now start the software, select the parallel port you are using and enter the distance between the photoelectric beams.

4. Click on the *Enable Measurement* button. The two green lights within the program display should light up. If not, make sure that the transmitters are sending the light to the receivers and that the receivers are able to receive the light.

5. Once the green lights are on, you're ready to go.

6. As soon as the first and then the second photoelectric beam are interrupted, the software responds with the computed speed.

When using the circuit, make certain that the distance between the photoelectric beams is not too small and not too great. Also, the time period between the triggering of the two photoelectric beams should not be too small nor too great.

Unfavorable values can cause the results to take on extreme proportions, which the software won't be able to convert. Use the circuit several times for testing and to gain experience. In this way you will learn how to use the circuit correctly to receive the desired results.

# Chapter 22:

# The Electronic Combination Lock

# Chapter 22: The Electronic Combination Lock

You've seen this circuit in *Mission Impossible* and James Bond films. An agent accesses a secure area by entering a number code on a touchpad, which disables an electronic lock. The circuit in this chapter takes this concept and converts it so that at the end of the chapter you will have an electronic combination lock. You can also use the circuit for other purposes.

You will build the circuit on half of a standard European ("pre-perfed") circuit board and attach it to the basic board you created in Chapter 9.

In addition, you will connect a *keypad* to the circuit that will allow you to input the numbers.

## The Circuit And Its Function

It's easy to understand how the circuit works. You input number combinations into the keypad attached to the circuit.

The circuit transfers the input to the computer through the parallel port. Once there, the software program evaluates the input, and if the input matches the previously specified number sequence, a signal is passed back to the circuit. The signal activates a relay, closing a circuit contact. Using this contact, you can enable an electronic door lock or a switch on some other appliance.

The circuit has two light-emitting diodes which indicate whether the number combination was entered correctly.

## Building The Circuit

The necessary components for the actual circuit can be quickly soldered to the board, since we only need a few components. However, there is an extra task to be performed. This has to do with the keypad, connected to the circuit by a flat ribbon cable. We'll tell you more about this task later.

First, build the add-on board. Follow the drawing in the next illustration.

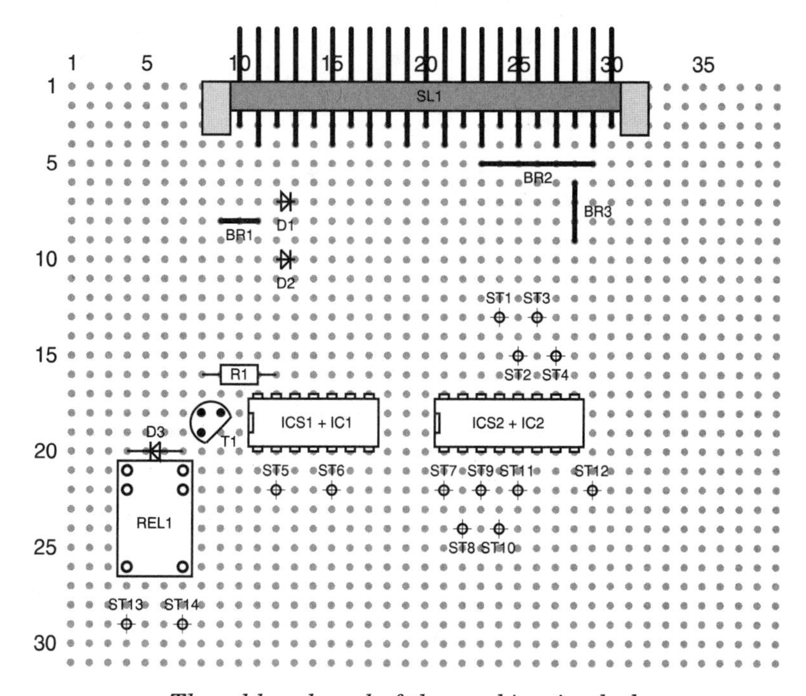


*The add-on board of the combination lock*

### Parts list

You should be able to solder the components to the board without any problems.

| Parts List | | |
|---|---|---|
| Label | Name | Component type |
| BR1 - BR3 | Conducting path bridges | Plastic-coated copper wire |
| ST1 - ST14 | Edge connectors (male) | Pin connection |
| R1 | 2.2 kΩ | Resistor |
| T1 | BC546 | Transistor |
| D1 | LED 5 mm (green) | Light-emitting diode |
| D2 | LED 5 mm (red) | Light-emitting diode |
| D3 | 1N4007 | Diode |
| REL1 | small relay 5 V, 1-pin UM | Relay |
| IC1 | SN7402 | NOR Gate |
| IC2 | SN74LS147 | Decimal-to-BCD Decoder |
| ICS1 | IC socket 14 pin | IC socket |
| ICS2 | IC socket 16 pin | IC socket |
| SL1 | edgeboard connection 21 pin | Pin connection |

The components are standard electronic elements. Note that the relay works with a rated voltage of 5 V. The coil resistance of the relay should not exceed 125 W.

Keep in mind that the relay you are using has limitations. You can only switch the amount of current or voltage for which the relay contacts have been designed.

## The Foil Side Of The Board

Refer to the following diagram for information on the conducting paths on the rear side of the board.

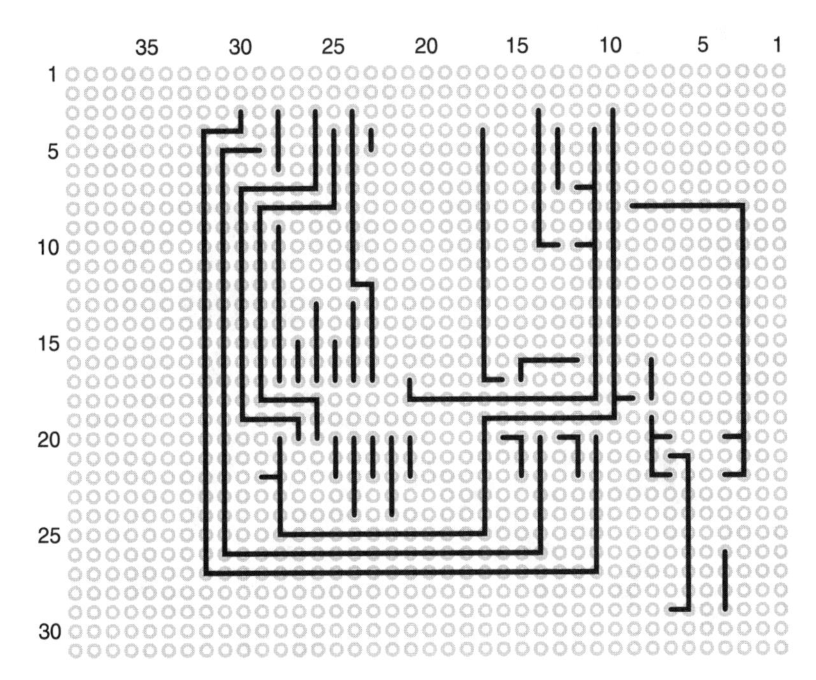


*The conducting paths of the combination lock*

After you finish making the connections between the components, the circuit should look like the following illustration.

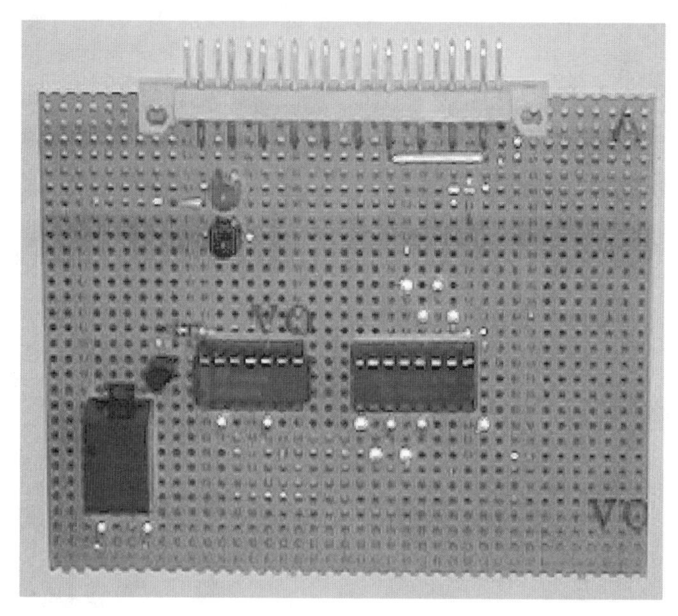

*The add-on board for the combination lock*

## Additional Work On The Circuit

After building the add-on circuit, you can turn your attention to the push-button keypad that you will use to enter the combination. You connect it to the pins of the circuit using a flat ribbon cable.

The keypad has 13 connections that include one connector which is connected to all the keys along with twelve separate contacts, one for each key. The keypad looks like this:

*The keypad*
*(shown here with a flat ribbon cable soldered on)*

Follow these steps to connect the keypad to the circuit.

1. Cut a piece of flat ribbon cable (13 conductors) to the desired length. Make sure it's not too short. On the other hand, don't let it exceed a length of 1 m.
2. Label the wire you solder to the common connector pin 1. You will recognize this contact by the fact that it is set off a bit from the other contacts.
3. At the other end of the flat ribbon cable, solder a male edge connector to each wire.
4. Finally, plug the wires into the matching female edge connectors of the circuit. Observe the following arrangement:

| ST12 | Wire 1 (common ground) | ST11 | Wire 8 (Key 5) |
|---|---|---|---|
| ST4 | Wire 3 (Key 7) | ST10 | Wire 9 (Key 2) |
| ST3 | Wire 4 (Key 4) | ST6 | Wire 10 (Key #) |
| ST2 | Wire 5 (Key 1) | ST9 | Wire 11 (Key 9) |
| ST5 | Wire 6 (Key 0) | ST8 | Wire 12 (Key 6) |
| ST1 | Wire 7 (Key 8) | ST7 | Wire 13 (Key 3) |

After you connect the keypad, you are finished building the circuit.

It's up to you to decide what to use the circuit for. Use the closing contact which is accessible via female edge connectors ST13 and ST14, to control whatever external device you want.

## Using The Circuit With The Companion CD-ROM

The software program analyzes the input from the keypad. This is located on the companion CD-ROM in the *Chap_22* directory. You can modify this software if necessary. The Visual Basic source code files are located in the *Source* directory.

When you start the software, you will see the following screen:

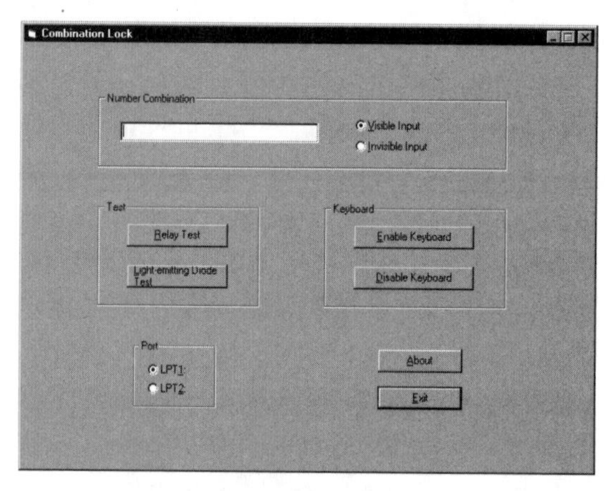


*The software program for evaluating the input*

You must specify a *Number Combination* in the software program so that the circuit can perform its task. The relay will only respond when you enter this number sequence using the keypad of the circuit.

Enter your numbers in the input box of the *Number Combination* area. You also have the option of making your input "hidden." As soon as you enter the numbers here, they are considered a combination.

You have to set the correct option for the port you are using so that the software can accept the input from the keypad. Choose either LPT1 or LPT2, depending on which port you are using.

After selecting the port and entering a number combination, enable the keypad by clicking the *Enable Keypad* button. The red light-emitting diode on the circuit will then light up. You can begin inputting numbers from the push-button keypad.

As soon as the correct number combination has been entered, the software program reacts and activates the relay for approximately five seconds. At the same time, the red light-emitting diode goes out, and the green one is enabled.

To test the function of the relay and the light-emitting diodes without having to enter the right key combination, click the *Relay Test* and *Light-emitting Diode Test* buttons in the software program.

### The extra keys [#] and [*]

The keypad has ten digits and two additional keys. These are the asterisk '*' key and the pound '#' key. The [*] key doesn't have any special function and is not connected. However, as soon as you press the [#] key, the entire number sequence you have entered to that point is deleted. This way you can correct to any mistakes you make entering the numbers.

## Using The Circuit

The following steps give you instructions on operating the circuit.

1. Make certain that the keypad is correctly plugged into the add-on board.
2. Plug the add-on board with the edgeboard connection into the socket strip of the basic circuit.
3. Connect the basic circuit to the parallel port of your computer.

4. Attach the external power supply to the basic circuit. Set the power supply to 9 volts and check the polarity. Then plug the power supply into the socket.
5. Start the installed software.
6. Set the *Port* option to LPT1 or LPT2.
7. Specify a key combination and then click on the *Enable Keypad* button.
8. The software accepts all the input of the circuit and evaluates it.
9. If necessary, you can check the function of the relay by clicking on the *Relay Test* button.

You can have the circuit perform different actions using the closing contact of the relay. Female edge connectors ST13 and ST14 are available for this purpose. Keep in mind the relay's circuit-breaking capacity, so that relay is not destroyed. You can use the relay to switch on an electric door opener, the light of a model train or a small motor. Use your imagination (within certain limitations).

# Chapter 23:

# PC Controlled Switching Turntable

# Chapter 23: PC Controlled Switching Turntable

The idea of the circuit which we'll talk about in this chapter is using the computer to control a motor. The circuit doesn't just switch the motor on and off, but rather, it controls the motor in such a way that it moves itself around a specified angle.

You cannot use a normal motor. A specific kind of motor is necessary for this purpose, called a *stepper motor*. This type of motor is able to execute defined angular movements by specific control signals. We'll tell you more about this concept later in the chapter.

You will build the control system of the motor on an add-on board in conjunction with the basic circuit you created in Chapter 9. Once again, half of a standard European ("pre-perfed") circuit board will suffice.

The add-on board, controlled by a software program, will send the appropriate signals to the stepper motor through control lines.

## The Circuit And Its Function

The circuit is designed to achieve the following function. Using a software program running on the computer, you can specify by which angle or up to which position the motor spindle is to turn. You can also influence the direction of rotation of the spindle.

When you give the command for executing the default action, then the software will send the necessary information to the circuit via the parallel port. The circuit, in turn, activates the stepper motor in such a way that the desired movement of the motor spindle is executed, thus taking the desired position.

As a result, the circuit is able to put a turntable driven by the motor spindle, e.g., made of wood, with a diameter of about 20 to 30 cm into rotation, so that it can take up the desired positions.

There's something else to say about the turntable. The idea of a computer controlled switching turntable is just one use for the stepper motor. Our idea has to do with a model train.

The disk to be turned should be integrated in a solidly attached turning mechanism provided with a driving spindle. The driving spindle is then connected to the spindle of the stepper motor with a drive band, e.g., with a rubber band. The drive of the stepper motor causes the disk to rotate.

The turntable must be positioned horizontally, so that on the upper side of the disk a track element of the model train can be attached. The entire structure is then integrated into an existing train system, so that a train can travel to the turntable.

Next, using the software/circuit combination, move the turntable by 180°, for example, so that the train can then travel back onto the track in the opposite direction.

Turning a locomotive around is just one idea. You could have sidetracks or other tracks on the turntable, so that a train could continue moving on predetermined tracks.

Building this turntable is time-consuming, and you'll have to invest lots of skill and inventiveness to have this extension for your model train function properly. Stores that sell model trains should have lots of helpful hints, e.g., how to provide the spindles with driving belts or corresponding locking devices.

Since both the size of the turntable and the integration option of this model train extension depend to a great extent on the existing model train, we cannot give any concrete specifications.

We'll briefly discuss two additional uses of this stepper motor circuit. The stepper motor, through its high precision, is well suited to act as the motor of a high lift forklift. That is, the motor of a model high lift forklift, e.g., built with Lego bricks. You can control the motor in such a way that it moves a hoisting mechanism to a specific position on the shelf.

The accuracy and control of the stepper motor can also be used for boring or milling. You can use the stepper motor to move the boring or milling table selectively and occupy specific positions.

You can achieve all these ideas using the stepper motor control system of this chapter. However, the control system is only part of the total structure. For the total structure, you will have to perform a series of additional, partially mechanical tasks that you must first carefully plan and prepare. The sky's the limit.

## Building The Circuit

The add-on board contains only a few components that are necessary for controlling the stepper motor. There are four small relays, with a diode for each relay, plus one IC chip.

Place the components on the board as shown in the following drawing.

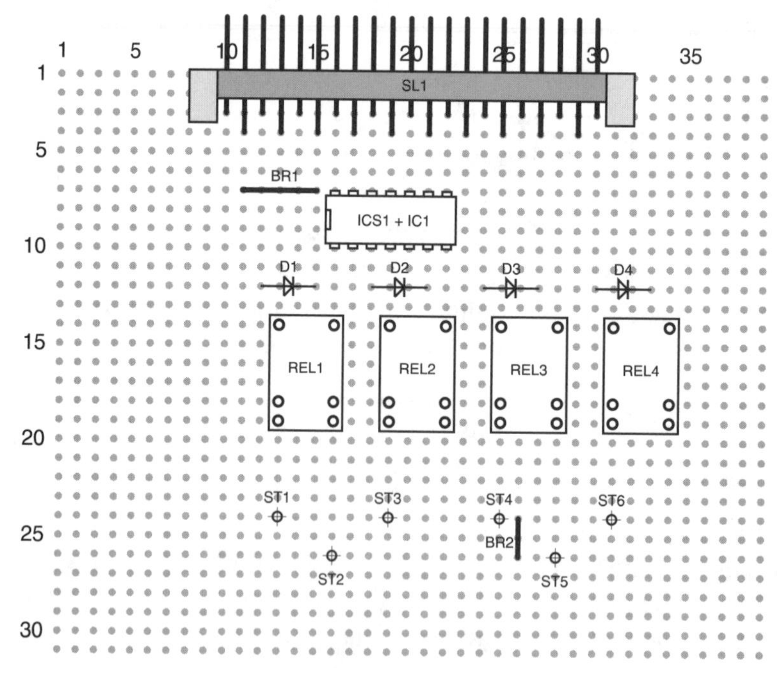


*The layout of the components on the add-on board*

When you build the board, make sure that the diode or band are soldered on properly. The marking of the diode consists of a colored circle on one end of the component. It marks the cathode of the diode, that is, the negative end of the component.

## Parts list

Refer to the parts list in the following table to determine which components are used in this chapter.

| Parts List | | |
|---|---|---|
| Label | Name | Component type |
| BR1 - BR2 | Conducting path bridges | Plastic-coated copper wire |
| ST1 - ST6 | Edge connector (male) | Pin connection |
| IC1 | SN7404 | Inverter |
| ICS1 | IC socket 14 pin | IC socket |
| D1 - D4 | 1N4001 | Diode |
| REL1 REL4 | Small relay 5 V, 1 pin UM | Relay |
| SL1 | Edgeboard connection 21 pin | Pin connection |

You are familiar with the components from the previous circuits of this book. Make certain that the relay works with a rated voltage of 5 V. The coil resistance of the relay should not exceed 125 W.

Since the relay controls the control signals for the stepper motor, do not exceed the limitations of the relay. The only time you can exceed these limitations is if you are using a large stepper motor. Then you would have to replace the relays with more powerful ones.

# The Foil Side Of The Board

Refer to the next drawing to wire the relays and diodes with the necessary connections.

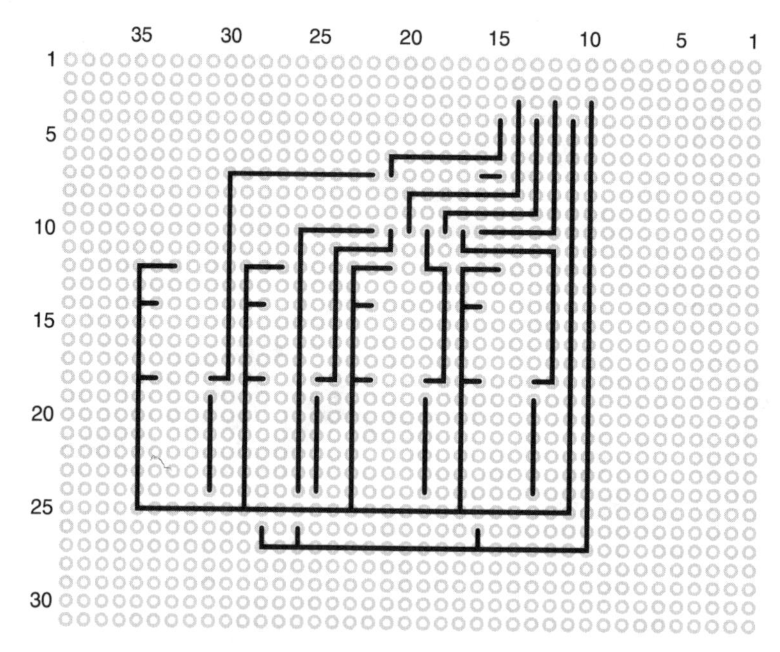


*The necessary conducting paths*

Here's what the add-on board will look like after you build it:

*The add-on board for the switching turntable*

## Additional Work On The Circuit

As you can easily gather, the circuit doesn't yet contain the stepper motor. You'll have to connect the stepper motor to the add-on board using several cables.

However, when doing this, there are a few items to take into consideration, since a stepper motor, in contrast to a normal motor, has several wires. You'll have to connect these to the contacts of the circuit.

Also, there are different types of stepper motors, which have to be connected and controlled differently. That is why we are going to give you a technical description of the stepper motor necessary for our circuit now.

For this circuit, you have to use a 2-phase belt unipolar stepper motor with 5 V operating voltage.

To have the stepper motor work properly with the software, the motor should execute 200 steps per turn, which corresponds to a step angle of 1.8°.

**Important Tip**

As we mentioned earlier, the spindle of the stepper motor drives the spindle of the turntable. If the spindles differs in diameter, the turntable will not move in 1.8° angular steps. Instead it will move based on the ratio of the step differential.

Using the following formula, you can calculate the angle of rotation of the turntable in relation to one step of the stepper motor:

```
Angle of rotation for turntable = 1.8°*(Stepper motor spindle
diameter/Turntable spindle diameter)
```

An example: If the stepper motor spindle diameter is 1 mm and the turntable spindle diameter is 20 mm, the following calculation results:

```
Angle of rotation for turntable = 1.8°*(1 mm/20 mm)
```

That corresponds to a turntable angle of rotation of 0.09°. That means that 1000 steps of the stepper motor (at 1.8°) on the turntable equal a rotation of 90°.

A 2-phase belt unipolar stepper motor has six control lines, so the motor has the following appearance.

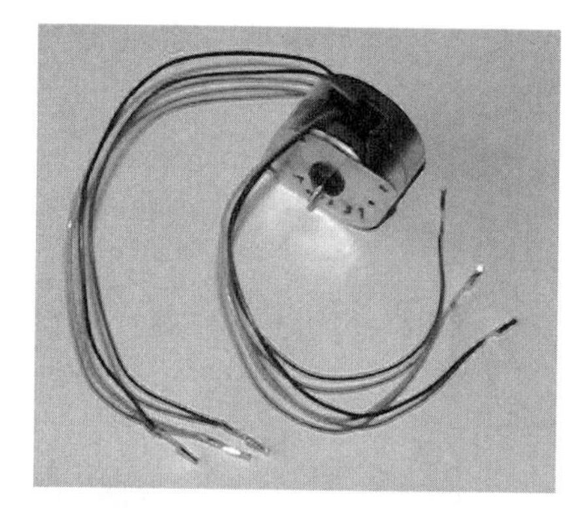

*A stepper motor with six control lines*

To connect the lines of the stepper motor to the circuit. There will be one problem. You have to find the correct lines and connect them with the contacts.

The stepper motor has two coils, each of which have what is called a *middle tap*. This must be set at ground, i.e., to 0 V. Each coil contains two additional wires, which have to be correctly connected to the contacts of the circuit. Otherwise the circuit won't be able to control the motor correctly.

Find the corresponding wires on your motor using the manufacturer's documentation and assign them to the wires on the drawing. In cases of doubt, have the dealer describe the wire assignment.

After that, proceed as follows:

1. The wires coming out of the motor aren't very long. Carefully extend them out until they reach from where you are using the motor to the circuit.
2. Strip the ends of the six cables and solder a male edge connector to each one.
3. Find the two wires for the middle tap of the two coils in the motor. Plug these into female edge connectors ST2 and ST5.
4. Plug wire 1 and wire 3 of coil 1 into female edge connectors ST1 and ST3.

5. Do the same with wires 4 and 6 of coil 2, connecting them to female edge connectors ST4 and ST6.

If you connected the wires properly, the software will control the motor correctly. The next section tells you how to check your wire connections.

## Using The Circuit With The Companion CD-ROM

As you have already learned, the software for this chapter's circuit has an important task. The software controls the stepper motor so that the motor makes the desired movement.

The software is located on the companion CD-ROM in the *Chap_23* directory. After installing the software you are ready to start the software program.

If you are planning to use the stepper motor control (software program) differently, you can make changes to the existing software. You will find the Visual Basic source code files in the *Source* directory for this software.

The software program for the computer controlled switching turntable offers the various options described in the following paragraphs.

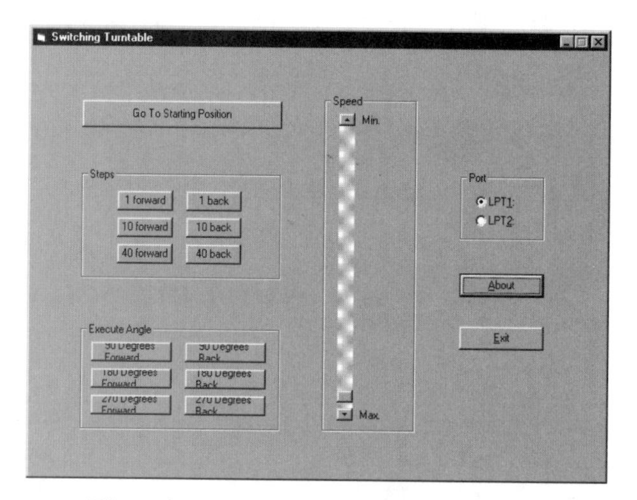


*The software for the switching turntable*

As you know, you need to select the correct port you are using, either LPT1 or LPT2. If you make the incorrect setting for *Port*, the software program will not transmit the signals to the circuit.

You will find a number of buttons displayed on the screen by the software program. First, let's talk about the *Go To Starting Position* button. As soon as you click this button, the spindle of the motor moves back to the starting position it had at the start of the program. This button gives you the opportunity of returning to the starting position from any position.

The *Steps* area contains six buttons, which allow you to move the motor forward or backward by a specific number of steps. Click the individual buttons to experiment with their effects. If the spindle moves in accordance with the button you clicked, it means you connected everything properly. If not, you will have to rearrange the control lines of the stepper motor to the circuit. When doing this, keep in mind that the terms forward and back refer to movement in a clockwise motion, although this is actually false in terms of angle units. However, it's easier for you to understand this way. When testing, do at least eight steps in each direction, since one or two steps will not guarantee that everything is connected properly.

Once you have checked and verified the control system of the motor, you can turn your attention to the other elements of the software.

The *Execute Angle* area places buttons at your disposal which make movements of the motor possible at a specific angle value. After you click a button, the stepper motor will set itself in motion until it has completed the specified rotation.

A different speed of rotation is required for different uses and depending on the size of the model train. To allow you to regulate the speed, there is a slider within the *Speed* area. Depending on which direction you move the slider, the speed at which the stepper motor performs its rotation changes.

**Important Tip**

Since the speed of rotation depends on the performance of your PC, you need to set the speed for your particular configuration. Make sure you do not set the speed too high, since otherwise the motor cannot be correctly converted. This has to do with the inertia of the relay. You're better choosing a lower speed to ensure correct function.

The correct and meaningful use of the software depends to a great extent on the pragmatic values you gather with the software in conjunction with the circuit and on the design of the turntable you use. Thus, the speed of rotation is not only dependent on the size of the turntable, but rather, also depends on the gear ratio you select of the stepper motor spindle diameter to the turntable spindle diameter. The rotational change of the turntable per step also depends on this ratio (as you were able to see in the sample calculation previously).

It makes good sense to experiment a great deal until you find the correct settings for speed and steps.

## Using The Circuit

No special tasks are required for putting the circuit into operation. However, there are a few points to take into consideration, and a great deal of experience to be gathered so that you obtain the right results.

Follow these steps when operating the circuit:

1. Connect the stepper motor to the circuit, you can even test the functionality of the circuit without a turntable.
2. Next, plug the add-on board with the edgeboard connection into the socket strip of the basic circuit. Use the printer cable to connect the circuit to the computer.
3. You will also need the external power supply for the basic circuit. Make sure that the power supply is set to 9 volts, and check for correct polarity. Only then do you plug the power supply into the socket.
4. If the turntable is already connected to the motor, turn the turntable manually in the desired starting position if necessary.
5. Start up the installed software program.
6. Set the option for the port to LPT1 or LPT2.
7. If you didn't turn the turntable manually to the starting position, you can use the buttons for this purpose, e.g., *1 forward* or *1 back*.
8. Set the speed of rotation to the desired value using the slider.
9. Then specify the desired movement of the turntable, and start the rotation using the described buttons.

**Important Tip**

If your stepper motor doesn't have an angle of rotation of 1.8°, enabling the buttons in the Execute Angle area will give you incorrect results. Also, under certain conditions, the spindle of the motor won't go back to the correct starting position since a wrong angle may have been specified.

As we mentioned, the function of the stepper motor can also be used for different purposes. You may have to make changes to the software for other applications---

.

# Chapter 24:

# The Remote Control Traffic Light

# Chapter 24: The Remote Control Traffic Light

This chapter's circuit functions as a remotely controlled traffic light. It uses light-emitting diodes to depict traffic light signals that are controlled by a program running on your PC.

Since you will build the circuit as an add-on board to be used with the basic circuit you created in Chapter 9, half of a standard European ("pre-perfed") circuit board will suffice.

The arrangement of the light-emitting diodes on the board resembles four traffic lights at an intersection. You can use the circuit with your model train, for example.

## The Circuit And Its Function

Our remote control traffic light is similar to the traffic light system you see everyday at a busy intersection. The traffic light signals will run the same as a real traffic light. There are two modes to choose from in the program:

1. The normal traffic light function with alternating red, yellow and green light phases.

2. A blinking phase in which only the yellow signals go on and off.

Since the add-on board only uses the light-emitting diodes as components, you can easily integrate it into a model train and activate it through a cable from the board. This lets you build separate traffic lights with the diodes and use them realistically.

## Building The Circuit

You can build the circuit quickly and easily because it only requires a few components. Most of the work was finished when you built the basic circuit.

Get the basic circuit ready so that you can create and use the remote control traffic light.

You will need just twelve light-emitting diodes and one pin strip for the circuit. Get the layout of the components from the following drawing.

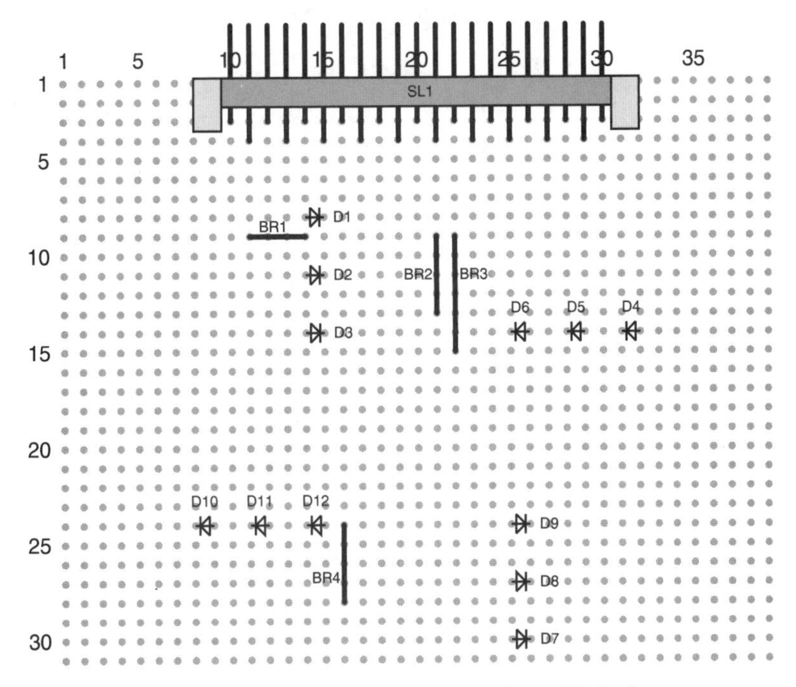


*The layout of the remote control traffic light*

Pay attention to the polarity of the light-emitting diodes. Remember the polarity changes from light to light.

## Parts list

Refer to the parts list in the following table to determine which components are used in this chapter.

| Parts List | | |
|---|---|---|
| Label | Name | Component type |
| BR1 - BR4 | Conducting path bridges | Plastic-coated copper wire |
| D1, D4, D7, D10 | 5-mm light-emitting diode (green) | Light-emitting diode |
| D2, D5, D8, D11 | 5-mm light-emitting diode (yellow) | Light-emitting diode |
| D3, D6, D9, D12 | 5-mm light-emitting diode (red) | Light-emitting diode |
| SL1 | Pin strip 21pin | Pin connection |

You can see in the parts list that you use 1/8 inch light-emitting diodes for the circuit. Since they also come in other sizes (from 1/8 to 3/4 inch), you can vary the size of the traffic light system. This can be advantageous for using the traffic light circuit in a variety of settings.

# The Foil Side Of The Board

To ensure the function of the circuit, you only have to wire up the twelve light-emitting diodes. Follow the conducting paths shown in the next drawing.

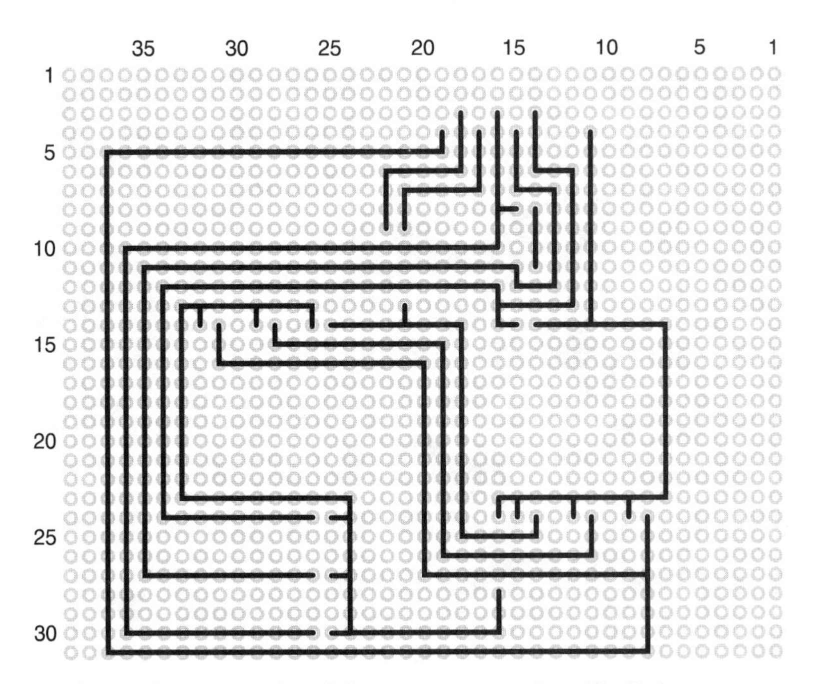


*The conducting paths of the remote control traffic light system*

Once you have installed the conducting paths, the add-on board is ready for the basic circuit.

The finished product will then look like the following illustration:

*The finished remote control traffic light*

If you look closely at the figure, you will see that we have drawn the course of the road on the board with a felt-tip pen. You can do this on your board too. That way you can tell which traffic signals are for which lanes.

## Using The Circuit With The Companion CD-ROM

This circuit works with a program, which controls the signals of the traffic light system.

Use the installation program on the CD-ROM to install the program. You will find it in the *Chap_24* directory. As usual, the Visual Basic source files are located in the *Source* directory.

After installation is complete, run the program.

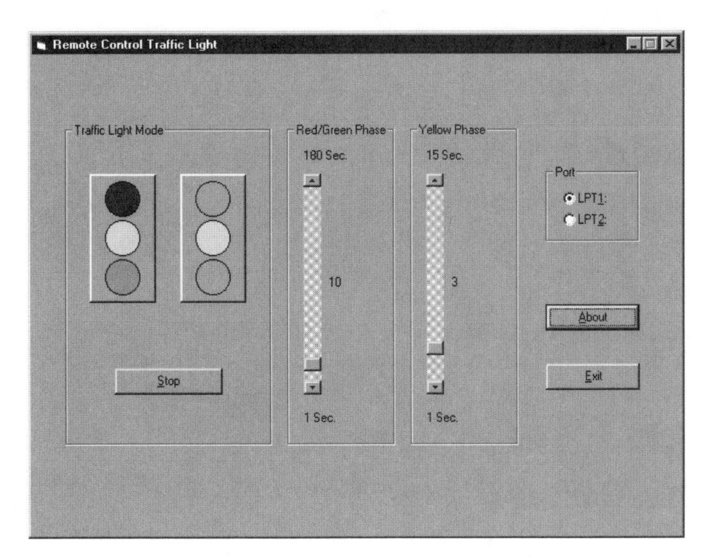


*Use this program to control the traffic light system*

As you can see, the circuit controls the traffic light system with only a few elements.

You can run the circuit from LPT1 or LPT2. Enable the correct option, LPT1 or LPT2, before starting the traffic light function.

Use one of the buttons from the *Traffic Light Mode* area to operate the traffic light system. There are two buttons here, which are responsible for the two modes we mentioned earlier.

If you click on the left traffic light button (all three signal lights are on), the circuit activates the regular traffic light system you are familiar with from street intersections.

If you click on the right traffic light button (only the yellow signal light is on), the traffic light system changes over to a blinking state.

You may change options during operation or after a stop phase.

The *Stop* button is available for switching off the traffic light system. After you click on this button, all the signals will remain dark.

Along with the choice of mode, you can also change the length of the individual signal phases. If the time interval in between the signal changes seems too long or too short, then move the two sliders in the desired direction. The signals will then change their state either more quickly or less often.

The slider for the *Red/Green Phase* influences the time span in which one street has a green light while the other one has a red light. The *Yellow Phase* slider controls the time span in which the yellow signal is on.

Experiment with the different possibilities, and find the best settings for your application.

## Using The Circuit

The following steps give you instructions on operating the circuit.

1. Plug the remote control traffic light add-on board with the pin strip into the socket strip of the basic circuit.
2. Connect the basic circuit to the parallel port of your computer using the printer cable.
3. Provide the basic circuit with an external power supply. Note the 9-volt setting of the power supply and the set polarity. Once everything is correct, plug the power supply into the socket.
4. After installing the program, start it up.
5. Enable the option for the port you are using, either LPT1 or LPT2.
6. Choose the desired operating mode by clicking on one of the traffic light buttons.
7. Adjust the time intervals between lights as necessary.

# Chapter 25:

# The Alarm System

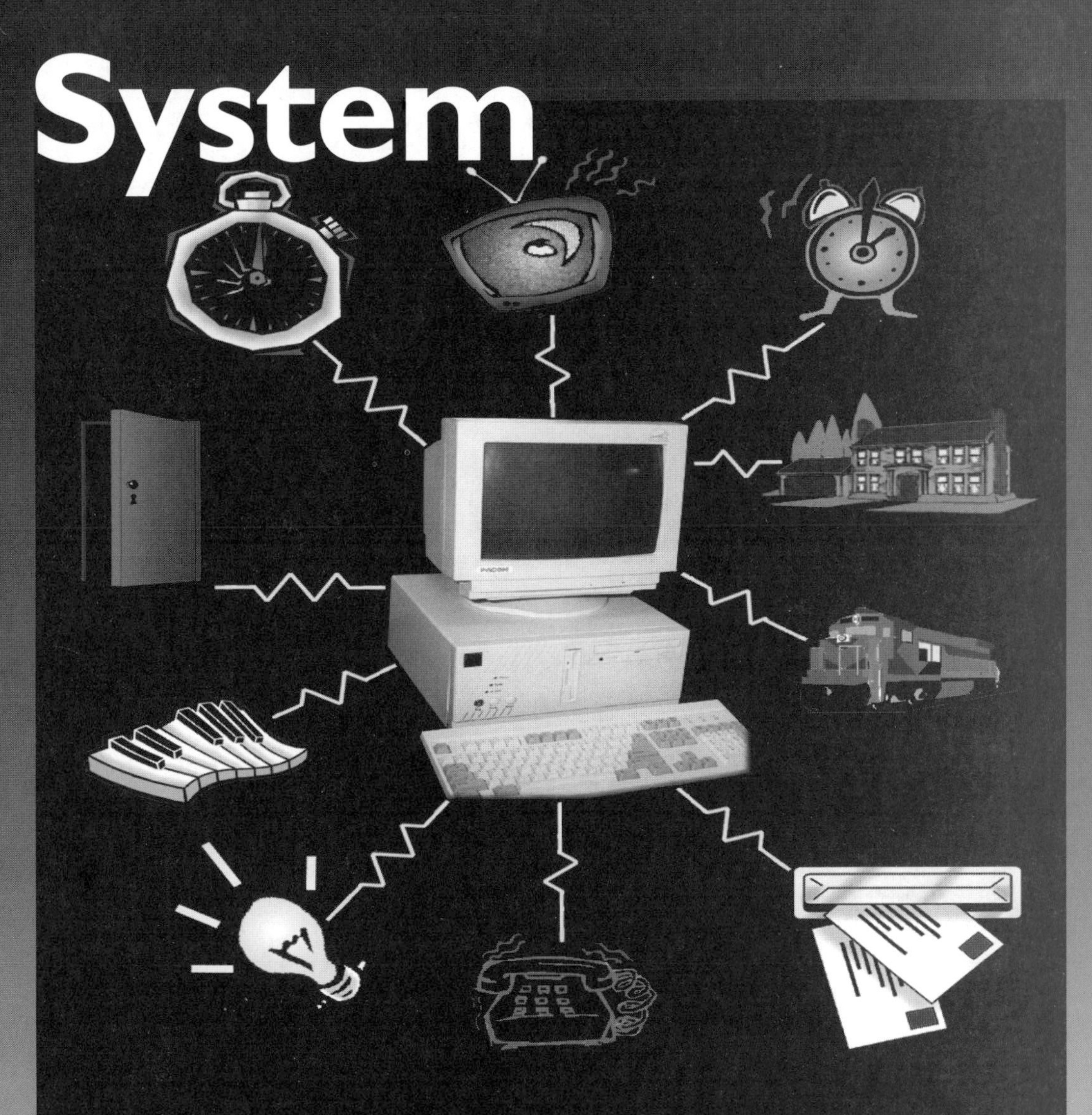

# Chapter 25: The Alarm System

The circuit you'll build in this chapter functions as an alarm, but don't expect a professional, high-performance alarm system. Instead, this circuit will furnish you with monitoring options that you can achieve with simple, reasonably priced resources.

This circuit passes information received from simple closing sensors to a software program running on your computer. The software program responds to these events with a blinking red light and a beep, informing you when one of the sensors has been activated.

As with the other circuits, this circuit exchanges data through the parallel port, working with the basic circuit from Chapter 9.

Since the main work with this circuit consists of attaching the sensors, the few required components will fit comfortably on half of a standard European ("pre-perfed") circuit board.

## The Circuit And Its Function

It's easy to understand how this alarm works. You can install up to three different closing contacts wherever you wish. For example, you could place them on windows, the door or under a doormat.

You connect the sensors' contacts to the circuit with two-conductor wire. If the state of one or more contacts changes, the circuit transmits the information to the computer. The software program on the computer responds with a display on the screen and an audible signal. In this way, you can quickly tell when something has changed on one of the sensors.

## Building The Circuit

Since the circuit is made up of just three resistors and some female edge connectors, you will be able to build it quickly.

The additional work will take a bit longer, but we'll tell you more about that later. First, refer to the following drawing for the locations of the components.

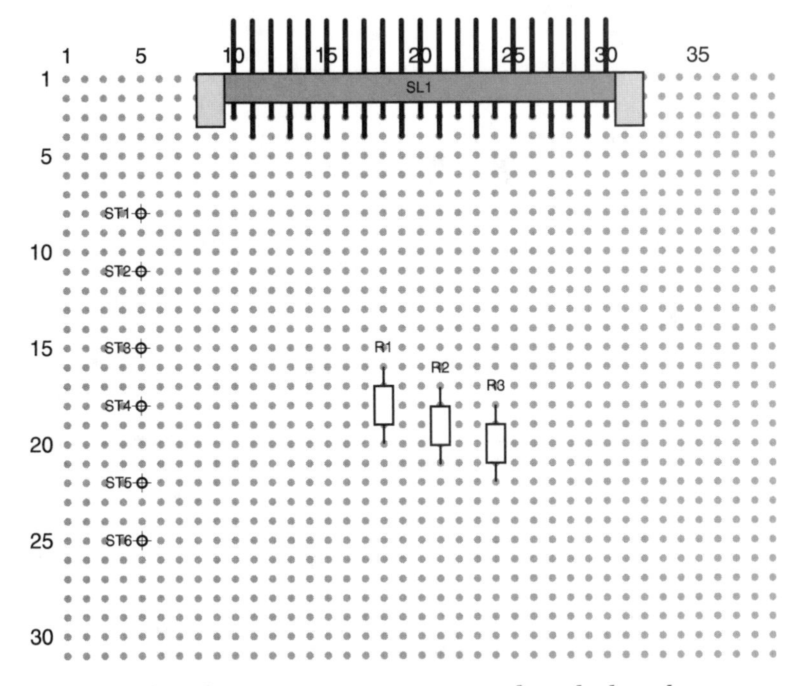


*How the components are arranged on the board*

### Parts list

Refer to the parts list in the following table to determine which resistors you need to use.

| Parts List | | |
|---|---|---|
| Label | Name | Component type |
| R1 - R3 | 1 kΩ | Resistor |
| ST1 - ST6 | Edge connector (female) | Pin connection |
| SL1 | Edgeboard connection 21pin | Pin connection |

You can build the circuit in a few minutes. There are no special features to consider, due to so few components.

## The Foil Side Of The Board

Since the circuit only contains a few components, only a few conducting paths need to be installed. The following drawing shows the conducting paths.

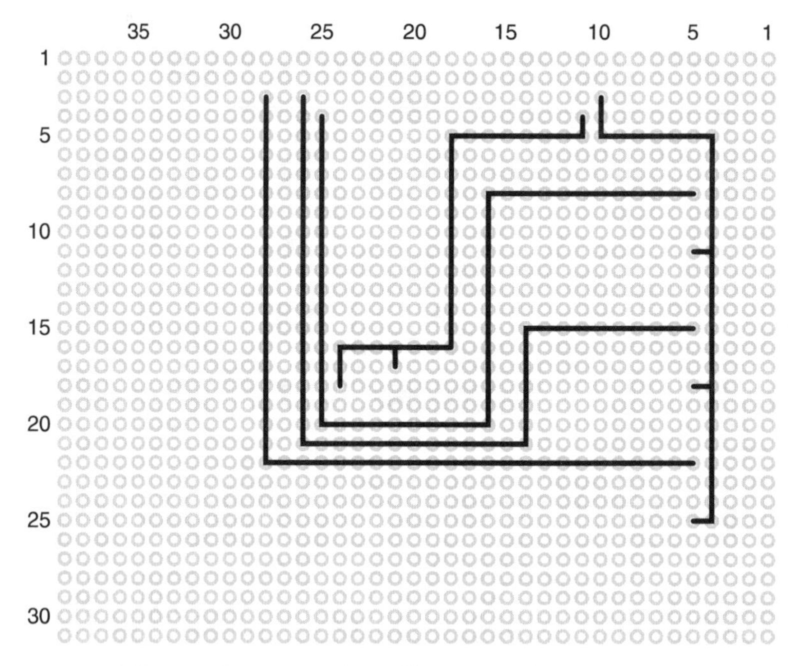


*A few conducting paths will connect the components*

Once you finish building the circuit, you are ready to turn your attention to the additional work.

*The add-on board of the alarm system*

## Additional Work On The Circuit

Proper use of the circuit depends on the sensors and their placement. But before we go into detail, you should prepare the connection to the computer and the external power supply.

Get the basic circuit ready, and plug the add-on board into the socket strip. The alarm system gets its power supply and contact to the parallel port of the computer through the basic circuit. Make sure that the basic circuit is provided with the necessary power supply and that the printer cable is connected.

Once you have completed these tasks, the circuit is ready to use. However, we still don't have the components that will feed the circuit with information: the sensors.

As you know, we are going to use simple *closing contacts*. It's up to you to choose which kind. For example, you could use a simple *microswitch*, available in various models in electronics shops.

Place the microswitch on a window frame, for example, so that opening the window triggers a switching event. The microswitch must trigger a closing function during the switching event, since the software is designed to react to this type of function.

You can also use other switches, such as *push buttons,* that you place in prominent areas, such as under the doormat. Many options are open to you for placing the sensors.

Once you have positioned your sensors, you need to connect them to the circuit. You will do this with a two-conductor cable. Each sensor must be connected to the circuit using such a cable.

Use the male edge connector to connect the cable with the circuit. Strip the two conductors of each cable and solder a male edge connector to each one of them. You will plug them into the female edge connectors of the circuit.

The circuit has six female edge connectors (three pairs) available for three sensors. Contact 1 is assigned to the pair ST1/ST2, contact 2 is assigned to ST3/ST4, and contact 3 is assigned to ST5/ST6. You don't have to worry about polarity when plugging in the two-conductor signal lines.

After plugging the signal lines into the appropriate female edge connectors of the circuit (remember, you don't have to use all three contacts), connect the other ends of the signal lines to the sensors. Which method you use to connect the signal lines to the sensors depends on the sensors. You could solder them, screw them on, etc.

# Using The Circuit With The Companion CD-ROM

The software for the alarm system is responsible for evaluating and responding to the data the sensors send to the circuit. As you know, you have to install the software before you can use it. The necessary files are on the companion CD-ROM in the *Chap_25* directory. The source code files are located in the *Source* subdirectory.

After loading the program, you are ready to start it up. When you do, the following screen appears.

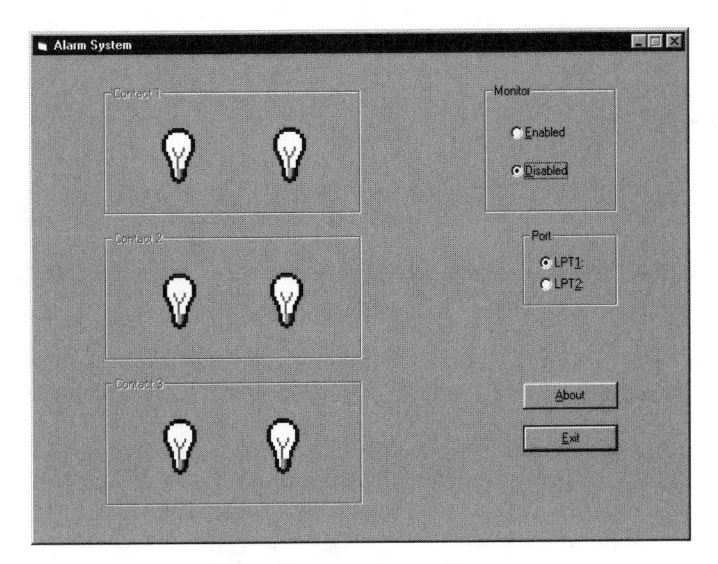


*The alarm system software*

Since the circuit controls the alarm system through the parallel port, this program also has the *Port* option. Choose the port from which you are operating the alarm system, either LPT1 or LPT2.

The software program has three similar areas, each one responsible for a sensor contact. The three areas labeled, *Contact 1*, *Contact 2* and *Contact 3,* are associated with a corresponding edge connector pair on the board.

Enable the alarm system using the radio buttons in the *Monitor* area. Click on the *Enabled* button to switch on the alarm. The contact areas become active, and a green light icon in each area signals its readiness. Click on the *Disabled* button to switch the alarm system off.

If the alarm system is active and a sensor triggers an alarm, the green light in the appropriate area goes out and is replaced by a blinking red light. The software program also outputs an audio signal. In this way you can easily tell when an alarm has gone off and which sensor caused it to go off.

## Using The Circuit

Operating the circuit does not require any special work. If you complete the following steps, nothing will stand in the way of using the alarm system.

1. Plug the add-on board into the basic circuit.
2. To guarantee that the circuit will transfer data to the computer, connect the basic circuit to the computer using a printer cable.
3. Make certain that the external power supply is near the basic circuit.
4. Place the sensors in the desired areas so that they are ready to be used. After that, connect the signal cable to the contact pairs on the circuit.
5. Start the software and make the necessary settings. First, set the correct option for the port you are using.
6. Start the monitor by clicking the *Enabled* button in the *Monitor* area.
7. Test the alarm system by intentionally triggering an alarm through one of the sensors.

Perform this last step with each sensor. If each of the sensors triggers an alarm, the system is functioning correctly, and you can use the circuit.

# Chapter 26:

# The Remote Control Train

# Chapter 26: The Remote Control Train

As you've probably assumed by the chapter title, this circuit is a control system that you can use with your model train. The circuit monitors a series of contacts. When these contacts are triggered, the program responds by activating corresponding relays. The relay contacts, in turn, can control external devices which you can use to influence the model train.

As with the other circuits, this circuit also exchanges data through the parallel port. Since you won't be using the basic circuit, all you need to prepare are the flat ribbon cable with the Centronics plug and the external power supply.

An important part of this circuit involves attaching the sensors. You will use special components, which we'll tell you more about later.

Prepare half a standard European ("pre-perfed") circuit board for building the circuit.

## The Circuit And Its Function

It's easy to understand how the circuit works. It monitors contacts that you can place and use anywhere. If one or more contacts are closed, the circuit and the program activate the corresponding relays, which control processes within the model train set.

The circuit has four input contacts for the sensors and three relays, which provide closing contacts for you. The two numbers differ because there isn't a constant one-to-on allocation here. While relay 1 responds to contact 1 and relay 2 responds to contact 2, activating the third relay requires that both contact 3 and contact 4 be active. As long as only one of these two contacts is activated, relay 3 won't react. So contacts 3 and 4 are logical AND operators. This arrangement lets you respond to specific situations more flexibly.

Two-conductor signal lines connect the sensors to the circuit. In the course of this chapter, you will learn about a special type of closing contact that is ideal for use in the model train set.

The program uses rectangle icons to update you on the current status of the circuit.

## Building The Circuit

The circuit consists entirely of components with which you are familiar, so you can easily build the circuit.

Because of the many female edge connectors in the circuit, a great deal of work is required to connect the input and output contacts to the model train.

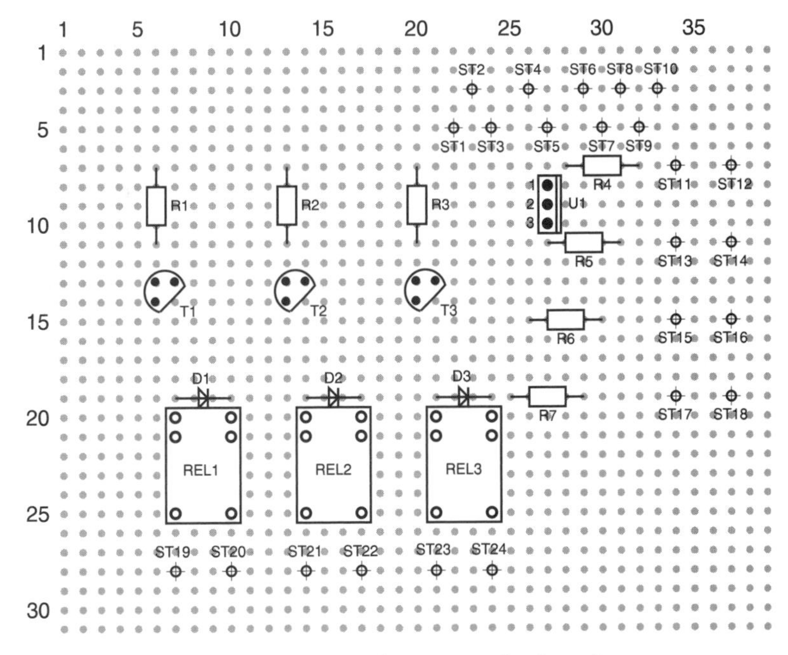


*The component layout on the board*

## Parts list

Refer to the parts list in the following table to determine which components are used in this chapter.

| Parts List | | |
|---|---|---|
| Label | Name | Component type |
| ST1 - ST24 | Edge connector (male) | Pin connection |
| R1 - R7 | 2.2 kΩ | Resistor |
| T1 - T3 | BC546 | Transistor |
| D1 - D3 | 1N4007 | Diode |
| REL1 - REL3 | Small relay 5 V, 1-pin UM | Relay |
| U1 | 7805 | Voltage regulator |

When building the circuit, make sure that the transistors and diodes are installed properly.

# The Foil Side Of The Board

Wire the circuit according to the conducting paths shown in the next drawing.

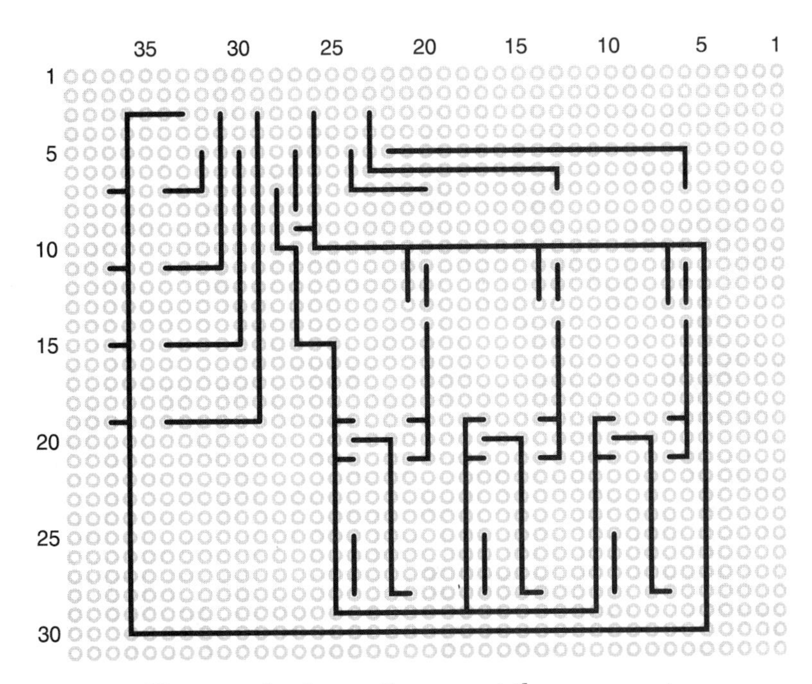


*These conducting paths connect the components*

After building the circuit, it will look like this:

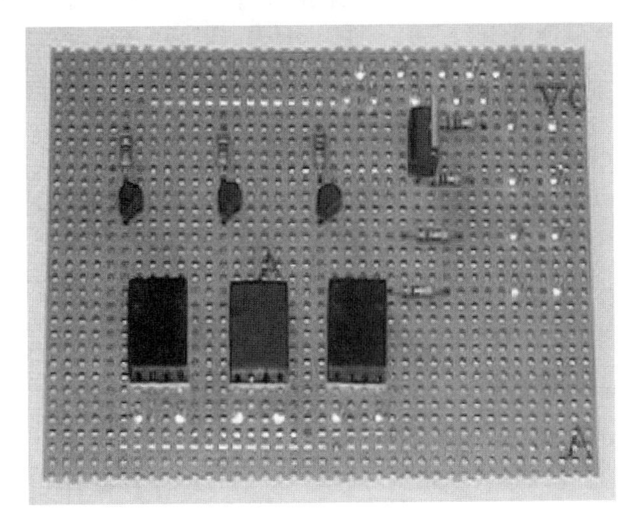

*The finished remote control train*

Now let's turn to the additional work, so you can make proper use of the circuit.

## Additional Work On The Circuit

You are already familiar with the first two tasks from other circuits. These involve the external power supply and the flat ribbon cable with the Centronics plug.

1. Use the external power supply with the adapter cable to provide the board with power. Plug the positive wire of the cable into female edge connector ST5. Plug the negative wire into female edge connector ST4. Now you've got power.

2. You have to connect eight wires from the flat ribbon cable to the female edge connectors of the circuit. The following table shows which wires join to which female edge connectors. If some of the wires of the flat ribbon cable still don't have male edge connectors soldered on, do this now.

| ST1 | Wire 3 | | ST7 | Wire 25 |
|---|---|---|---|---|
| ST2 | Wire 5 | | ST8 | Wire 23 |
| ST3 | Wire 7 | | ST9 | Wire 19 |
| ST6 | Wire 28 | | ST10 | Wire 2 |

Once you finish these tasks, the circuit is ready to use. However, you still need the sensors that feed the circuit with information. As you know, you can use simple *closing contacts*. It's up to you to choose which type. For example, you could use simple *microswitches*, which then trigger the contact.

Owners of model trains often want to have a contact react to a passing train. While you could also achieve this end with mechanical contacts, a more elegant solution involves *reed switches*.

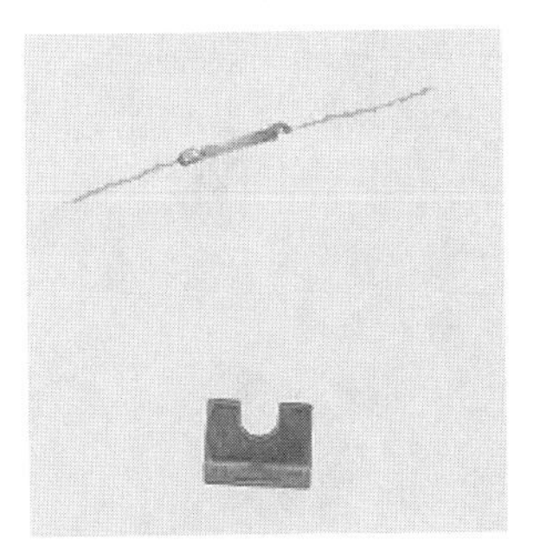

*The reed switches are often very small (a magnet is shown below the switch)*

In many cases, a reed switch consists of a small glass tube containing the opened contact. The tube is closed on both ends, with a contact wire sticking out on each side. You can connect these wires to the contacts of the circuit.

Now you're probably wondering how the contact gets closed, since it isn't accessible inside the glass tube. That's precisely what makes this type of contact so interesting for a model train. You can close the contact with the help of a magnet. As soon a magnet gets close enough to the glass tube, the contact closes.

Based on this, you can attach the glass tube to the railroad tracks and make a connection from the switch to the circuit by means of a two-conductor wire. Then attach magnets to the trains.

As soon as the train, and thus the magnet, passes over the glass tube, the contact closes and you can undertake the appropriate controls through the corresponding relay of the circuit.

The circuit has eight female edge connectors (four pairs) available for the four sensors. Edge connector pair ST11/ST12 is assigned to contact 1, edge connector pair ST13/ST14 is assigned to contact 2, edge connector pair ST15/ST16 is assigned to contact 3 and edge connector pair ST17/ST18 is assigned to contact 4. When plugging in the two-conductor signal lines, you don't need to worry about polarity.

When placing the sensors, see to it that contacts 3 and 4 are logical AND operators. You will only get useful function if these contacts perform combined tasks.

As soon as you have installed and connected the sensors, you can also connect cables to the output contacts of the relays and use them to control gates or similar objects. Use edge connector pairs ST19/ST20, ST21/ST22 and ST23/ST24.

## Using The Circuit With The Companion CD-ROM

The program for the remote control train circuit evaluates and responds to the data the sensors send to the circuit. However, you cannot use this program until you install it. The files are stored on the companion CD-ROM in the *Chap 26* directory. The source code files for the program are located in the *Source* subdirectory, in case you wish to modify the existing program.

Start the program to display the following window on your screen.

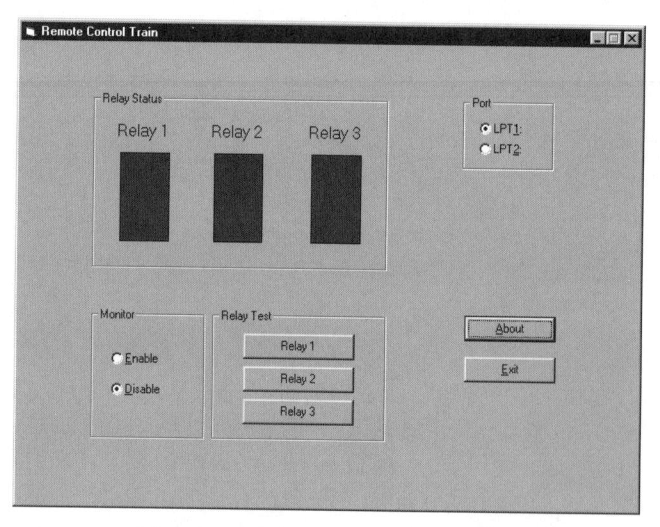


*The remote control train program*

You run this circuit from the parallel port, so set the appropriate option, either LPT1 or LPT2.

The *Relay Test* area has three buttons, one for each relay, allowing you to test the function of the relays separately from the input contacts.

As soon as you click on the buttons, you will see a change in the *Relay Status* area. This area has three rectangles, which display whether a relay is currently enabled or disabled. The rectangle is green for active relays, otherwise it is red.

To have the circuit run in conjunction with the program, click the *Enable* option in the *Monitor* area. Only then will the program evaluate the information from the input contacts and transmit the appropriate reactions to the relays.

## Using The Circuit

The following steps give you instructions on operating the circuit.

1. To make data exchange possible, connect the board to the computer using the flat ribbon cable and the printer cable.
2. Place the sensors in the desired areas so they are ready to use. Then connect the signal cable to the contact pairs on the circuit.Make certain that the external power supply is near the board and plugged into the socket.
3. Make certain that the external power supply is near the board and plugged into the socket.
4. Start the program and set the correct option for the port you are using.
5. Check the function of the relays by clicking on the appropriate buttons.
6. Start the monitor by clicking the *Enable* option in the *Monitor* area.
7. It's a good idea to intentionally trigger each of the sensors. When doing this, remember that the triggering must occur simultaneously with contacts 3 and 4, so that relay 3 becomes active.

Once you have run these tests successfully, the circuit is ready to use.

# Chapter 27:

# The Remote Control Socket

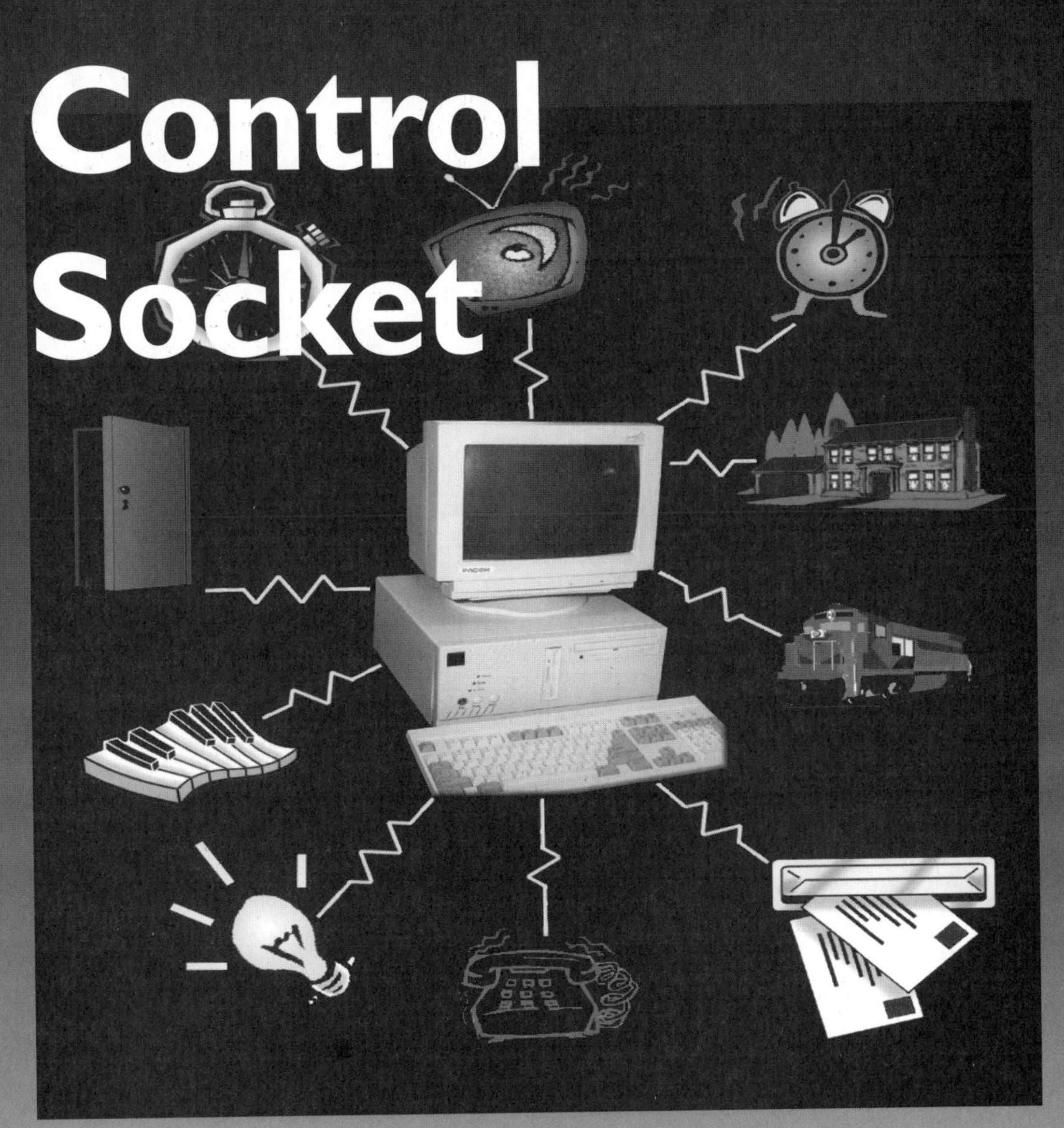

# Chapter 27: The Remote Control Socket

Building the circuit in this chapter will let you switch two sockets on or off. You can control electrical appliances from your PC that are plugged into the sockets. The program that operates the circuit provides the control options. The program transmits data through the parallel port of the computer to the circuit to enable or disable the relays, which control the supply of power to the sockets. Although you could also control other circuits connected to the relays, our circuit is designed to handle the load up to 220-volt appliances. **Make certain to read the safety notes at the end of this chapter.**

The circuit must be operated with the external power supply to work properly, particularly with regard to the circuit's relays. You can build the circuit on half of a standard European ("pre-perfed") circuit board.

## The Circuit And Its Function

The program lets you control the signal states of two of the eight data lines of the parallel port. The relays, in conjunction with the transistors, are connected to these data lines. When a data line becomes active, the appropriate relay switches. Then the circuit connected by the contacts of the relay can be switched on or off. In this case the switching off can be performed with the opening contact of the relay.

Besides the external power supply, the circuit will also need the flat ribbon cable with the Centronics plug for the parallel port of the computer. You'll plug the existing male edge connectors of this cable into the appropriate female edge connectors of the board.

## Building The Circuit

To ensure that all the components will fit on half of a standard European ("pre-perfed") circuit board, make certain to arrange them very tightly on the board. This especially applies to the relays, which are tall compared to the other components.

For this reason, begin by soldering the components that lie flat on the board. Worry about the relays and connecting terminals at the end. This will give you easier access to the smaller components.

Build the circuit as shown in the following drawing.

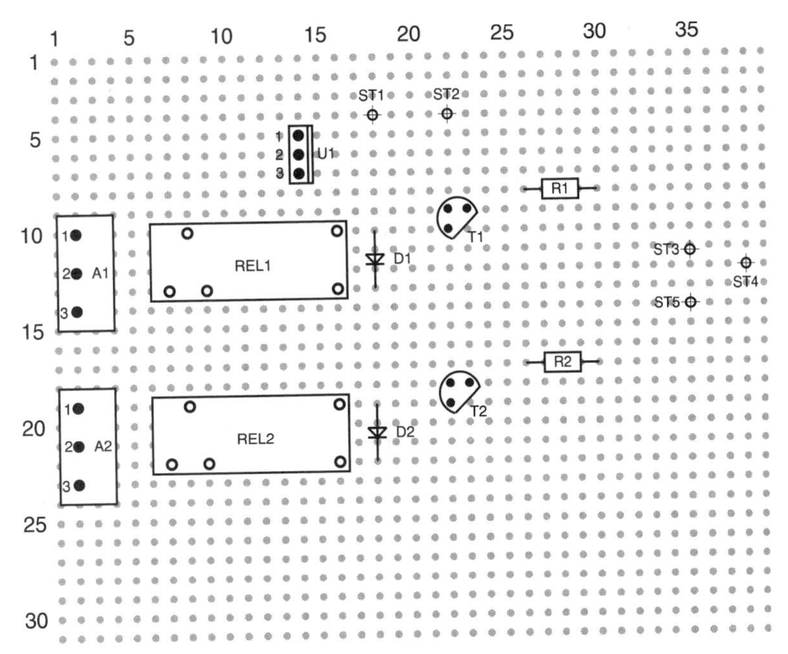


*The layout for the remote control socket*

### Parts list

Refer to the parts list in the following table to determine which components are used in this chapter. As you can see, the circuit requires only a few components.

| Parts List | | |
|---|---|---|
| Label | Name | Component type |
| R1 - R2 | 2.2 kΩ | Resistor |
| T1 - T2 | BC546B | Transistor |
| D1 - D2 | 1N4007 | Diode |
| REL1 - REL2 | 5-V power relay 1 x UM | Card relay |
| A1 - A2 | 3-pin connecting terminal | Connecting terminal |
| U1 | 7805 | Voltage regulator |
| ST1 - ST5 | Edge connectors (female) | Pin connection |

## Building the circuit

When working with the components, be sure to follow our advice about which components to solder first. Also, keep the following points in mind:

1. Once again, observe the polarity of the diodes.
2. Don't press the transistors down completely to the bottom of the board. Let them stick out a bit beyond the board. This way you'll be able to reuse them if you no longer require this circuit.
3. The relays are the same ones you used for the circuits in Chapters 15 and 19. Refer to these chapters for the appropriate specifications of the relays.
4. The connecting terminals have to be designed for mounting on a standard circuit board.

# The Foil Side of the Board

Refer to the following diagram for information on the conducting paths.

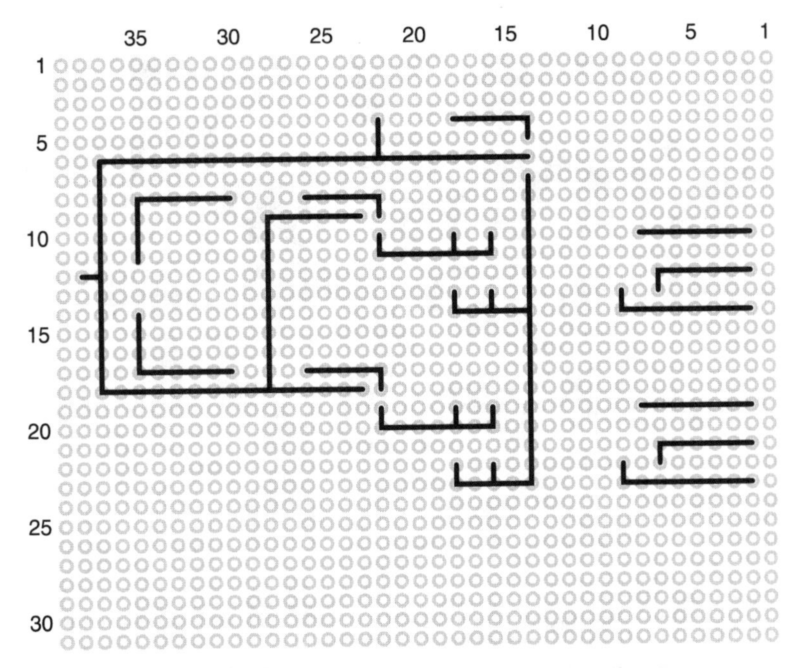


*The conducting paths for the remote control socket*

## Additional Work On The Circuit

Follow these steps to attach the external power supply and Centronics cable for the computer. The circuit is ready to use when you have completed them.

1. The two female edge connectors ST1 and ST2 are available to receive the external power supply. Connect the positive wire to female edge connector ST1 and the ground wire to female edge connector ST2.

2. Plug the flat ribbon cable with the Centronics plug to the indicated female edge connectors of the circuit. Refer to the table on the right for the proper arrangement of wires and connectors.

| ST4 | Wire 2 |
|---|---|
| ST3 | Wire 3 |
| ST5 | Wire 5 |

When done, your circuit will look like the following:

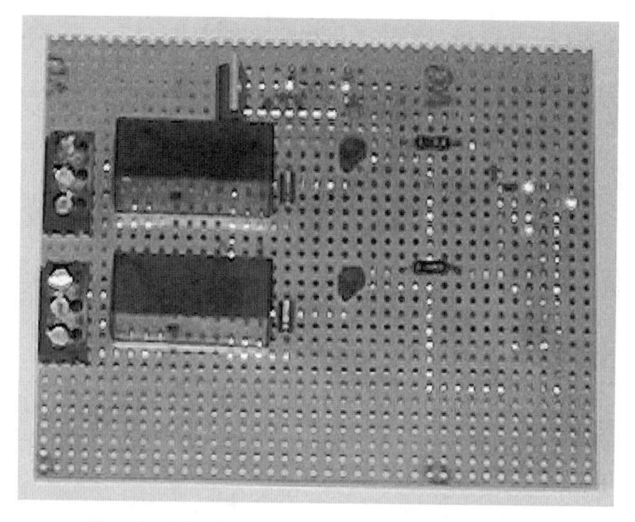

*The finished remote control socket circuit*

## Using The Circuit With The Companion CD-ROM

The program is responsible for an important part of the total function. The circuit only switches the relays on or off. The program furnishes you with the control.

The program is on the companion CD-ROM in the *Chap_27* directory. If you need additional functions for controlling the relays, you can implement them with the Visual Basic source code files, located in the *Source* subdirectory.

Now we'll explain the default functions.

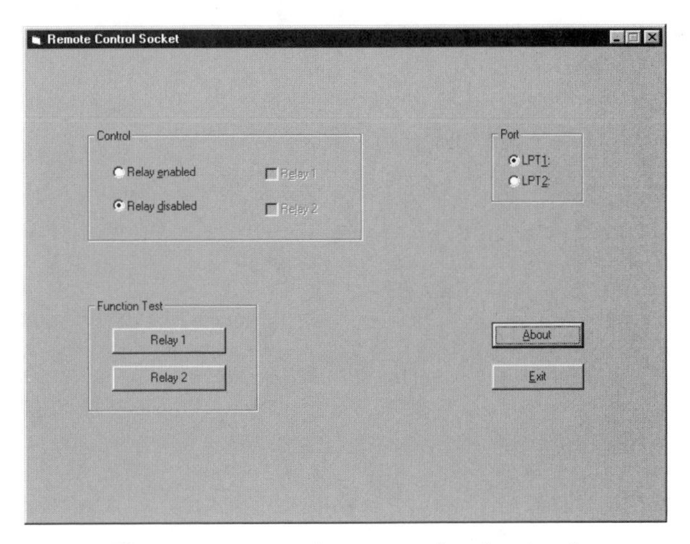


*The remote control program for the circuit*

This program lets you set the desired parallel port to either LPT1 or LPT2.

Basically, the control program for the relays is divided into two areas. The program has the *Function Test* area for checking the relays. When you click the *Relay 1* or *Relay 2* button, the program disables the set relay parameters and activivates the relay for approximately two seconds. Then it reverts to the original state of the relay.

You control the relays in the *Control* area, where you have the option of enabling or disabling the relays. Only when control is active can you switch the two relays on or off independently of each other using the *Relay 1* or *Relay 2* check boxes.

Using these functions you can control the circuit, and along with it, the sockets.

## Using The Circuit

The following steps give you instructions on operating the circuit. **Again, make certain to read the safety notes at the bottom of the next page.**

1. First, connect the circuit to the parallel port of the computer.
2. Next, connect the external power supply to the circuit, as explained earlier in the chapter. Plug the cable of the power supply into a normal socket.
3. Now turn your attention to the program. Run it and set the correct option for the port you are using (LPT1 or LPT2). Click on the buttons in the *Function Test* area to test the relays.
4. Now you are ready to control the circuit using the available functions.

Using the contacts of the relays or the connecting terminals, you can now connect the lines for controlling the sockets. Terminals 1 and 2 form an opening contact, while terminals 1 and 3 form a closing contact.

Before connecting and using these lines, be sure to read the following notes.

## Caution when using this circuit

The circuit is designed to allow you to control an alternating voltage of up to 220 volts using the relays. Since the contacts of the relays are freely accessible underneath the board, there is a safety risk.

You need to make sure that these conducting paths cannot be touched. Mount the circuit within a non-conducting case.

Since you want to control sockets with the switch, it's a good idea to integrate them into the case. That way you can install the control lines within the case, so that safety is guaranteed, even in regard to the socket connections.

**Important Tip**

In any event, consult with an expert and have an expert perform the necessary tasks. Under no circumstances should you connect a 220 volt line to the connecting terminals if the circuit is freely accessible.

# Part III: Circuits For Advanced Users

Part III talks about two more circuits. Their functions and designs are considerably more demanding than the circuits so far. These circuits cannot be built on a circuit board. They have conducting path layouts that you can use to have boards etched. We'll tell you more about that later.

Also, the circuits contain memory chips or controller chips which have to be programmed for use. The chapters will explain how to program them.

The circuits in Part III are obviously more labor-intensive and time-consuming to build. The methods you've used to now are not enough for these more complicated circuits. However, you don't have to be a professional electrician to build the circuits. You'll just have to perform tasks that take longer and involve more work.

As was the case with the previous circuits, the chapters are structured so that you get the required tasks in the correct sequence with the necessary explanations. This makes it possible for you to build working versions of the circuits.

# Chapter 28:

# The Burner Circuit

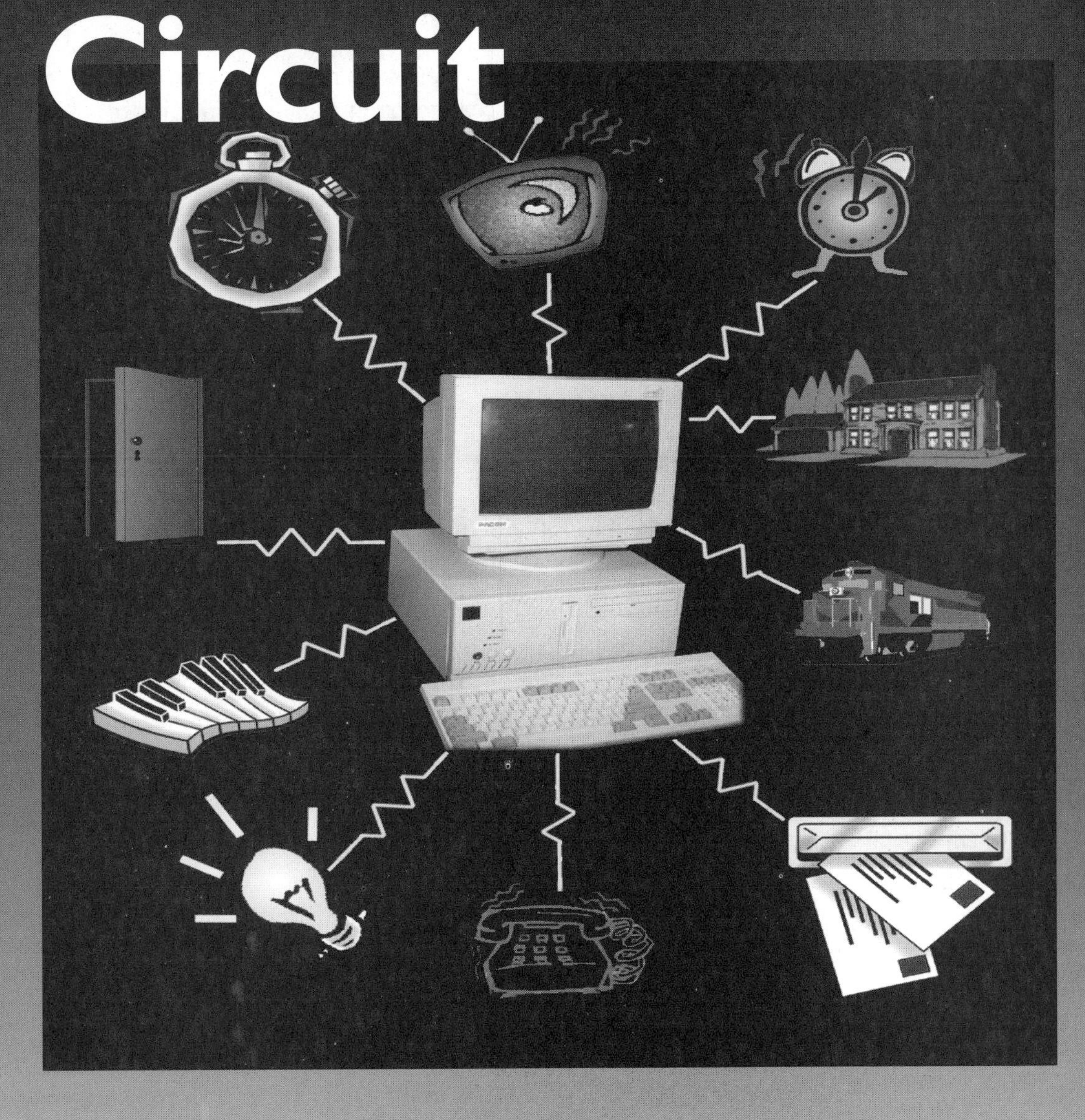

# Chapter 28: The Burner Circuit

Don't be put off by the name of this circuit—it has nothing to do with fire. The two circuits after this one use controller chips which possess a small internal memory space. This memory space can receive information that controls the function of the controller. Transferring this information into the memory range is referred to as "burning" because the information is burned into the chip. Even when the controller is no longer supplied with power, the information remains stored in the chip.

## The Circuit And Its Function

Now that you know what we mean by "burning," the term *burner circuit* is quite meaningful. The circuit has the sole task of placing specific information onto the controller chips of the circuits in this chapter and in Chapter 29.

The circuit consists of a standard European ("pre-perfed") circuit board with various chips. Among others, there is an IC socket in which the controller chips are inserted. A cable connects the circuit to the serial port of the computer, and an external power supply has to be connected to the board.

On the computer, the required information for the controller is provided in files contained on the companion CD-ROM. A program transfers this information to each controller chip.

After transferring the information, the burner has completed its task. The controller can be taken out of the IC socket and the circuit can be disconnected again.

## Building The Circuit

Building the burner is quite easy. It requires only a few reasonably priced components, so you can quickly build the circuit. Along with the components listed in the parts list, you will also need the external power supply with 9 volts of output voltage.

You can build the entire circuit on half of a standard European ("pre-perfed") circuit board. You are already familiar with the components from the circuits of Part II. Consult the following drawing for the layout of the components on the board.

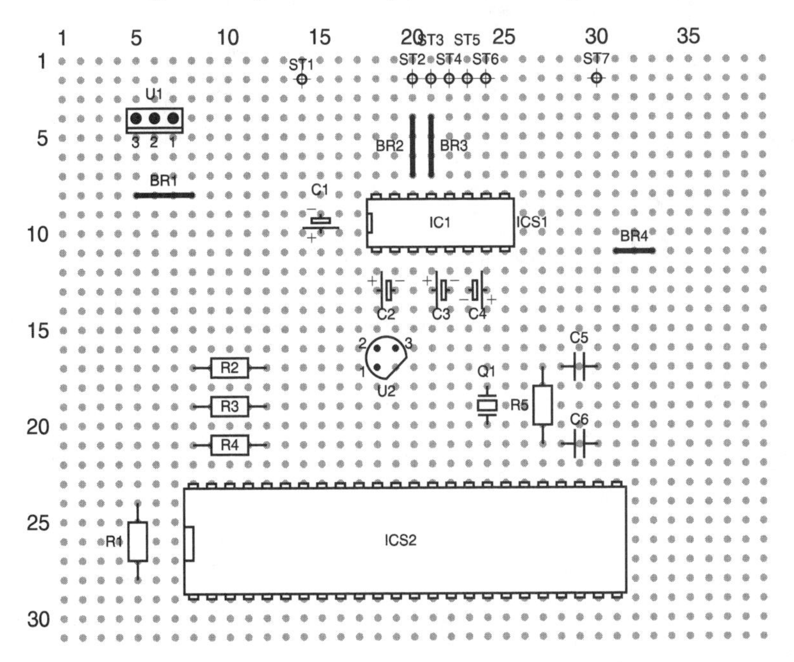


*The component side of the burner board*

### Parts list

Refer to the parts list in the following table to determine which components are used in this chapter.

| Parts List | | |
|---|---|---|
| Label | Name | Component type |
| BR1 - BR4 | Conducting path bridges | Plastic-coated copper wire |
| R1, R2, R4 | 4.7 kΩ | Resistor |
| R3 | 1 kΩ | Resistor |
| R5 | 1 MΩ | Resistor |
| C1, C4 | 10 µF | Condenser (Tantalum) |
| C2, C3 | 1 µF | Condenser (Tantalum) |
| C5, C6 | 1 pF | Condenser |
| U1 | 7805 | Voltage regulator |
| U2 | MC34064 | Reset chip |
| Q1 | 8.000 MHz | Quartz Mini |
| IC1 | IC MAX232 | IC |
| ICS1 | IC socket 16 pin | IC socket |
| ICS2 | IC socket 48 pin | IC socket |
| ST1 - ST5 | Edge connectors (female) | Pin connection |

### Building the circuit

There are no special features to remember when you build this circuit. After soldering the components into the specified positions on the board, you can begin making the required connections on the foil side of the board.

## The Foil Side of the Board

Refer to the following drawing for the proper connections of the components on the foil side of the board:

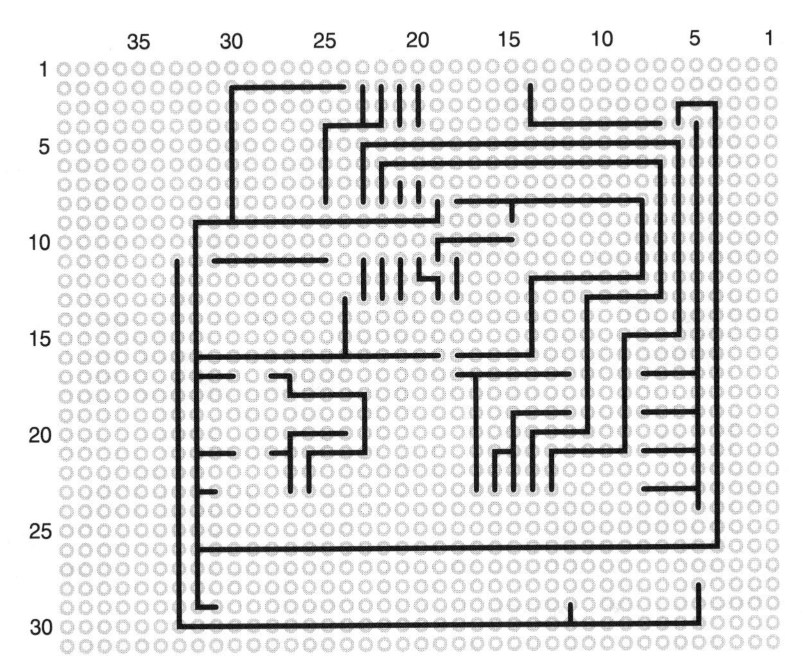


*The conducting paths of the burner board*

When you finish making the connections, a few additional tasks are necessary before the circuit will be ready to use.

## Additional Work On The Circuit

Notice that the board has five female edge connectors. You will use these to create the required connection to the serial port of your computer.

You will need the following materials to create this connection:

- approximately 20 to 40 inches of 5-conductor flat ribbon cable
- a 25-pin Sub-D connector (female, with case)
- 5 female edge connectors

**Important Tip**

If you furnish the connector with a case, consider providing tension relief for the cable.

## Performing the tasks

1. Once you have the materials, solder a female edge connector to each wire on one end of the flat ribbon cable. On the other end of the cable, create the connections to the following five pins. Mark wire 1 of the flat ribbon cable first, so that you can use the correct assignment.

```
Wire 1      to Pin 2
Wire 2      to Pin 3
Wire 3      to Pin 5
Wire 4      to Pin 6
Wire 5      to Pin 7
```

2. Once you have soldered the cable ends to the appropriate pins, the cable is ready to use. All you need to do now is plug the male edge connectors into the correct female edge connectors on the board.

```
Wire 1 to edge connector ST2
Wire 2 to edge connector ST3
Wire 3 to edge connector ST4
Wire 4 to edge connector ST5
Wire 5 to edge connector ST6
```

The circuit is finished once you have plugged the male connectors into the females.

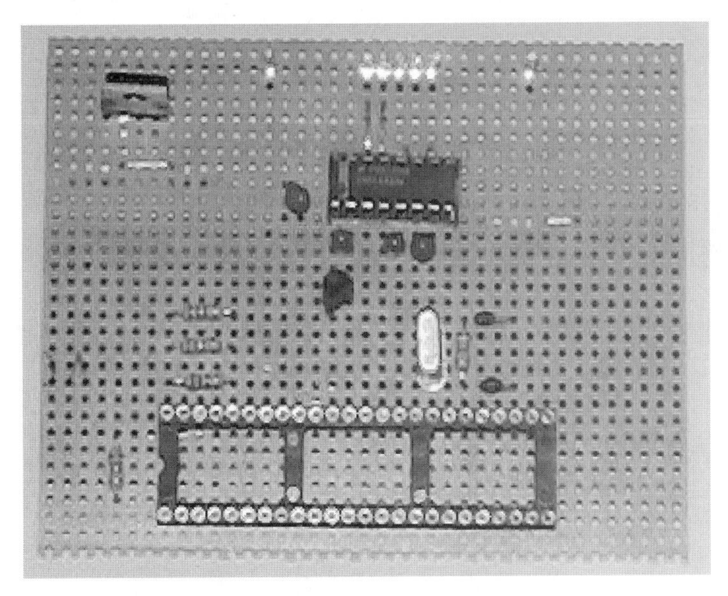

*The finished burner board*

## Using The Circuit With The Companion CD-ROM

The program for the burner circuit is located on the companion CD-ROM in the *chap_28* folder.

Here are two batch files and two files with the *.out* extension. The two batch files are practically identical. The only difference is that the file named *BURNER1.BAT* addresses the COM1 port, while the file named *BURNER2.BAT* addresses the COM2 port.

```
BURNER2.BAT
cls
@echo off
echo
```

```
echo •
•
echo • please connect burner to COM2 and switch on +5V, •
echo •
•
echo • then continue by pressing any key
•
echo •
•
echo
•••••••••••••••••••••••••••••••••••••••••••••••••••••••••
pause
cls
echo sending loader
mode com2: 12 n 8 1
copy /b loader.out com2:
mode com2: 30 n 8 1
echo Sending program and programming
echo ••••••••••••••••••••••••••
echo •                          •
echo •      please WAIT..      •
echo ••••••••••••••••••••••••••
copy /b io.out com2:
cls
echo FINISHED, 68HC11 is programmed
```

**Important Tip**

If you wish to use the circuit on a COM3 or COM4 port on your computer, simply copy the entire contents of the directory to your hard drive and edit one of the two batch files. Replace all character strings containing "COM1" or "COM2" with "COM3" or "COM4." Then save the modified file as BURNER3.BAT or BURNER4.BAT.

The two *.out* files contain data that must be sent to the controller. If you try to view the contents of the files with an editor, you will notice that they are not readable. The data is in a format that the controller can read.

Using the program doesn't require any special precautions. Simply start one of the two batch files at a DOS prompt. We'll tell you more in the next section.

## Using The Circuit

This circuit's task is to program the controller for the circuit in the next chapter. To do this, follow these steps.

1. Plug the circuit's connector into one of your PC's serial ports (as a rule, it should be COM2). Then plug the 68HC11A1 controller into the IC socket provided for it on the circuit. Make sure you insert the controller correctly.

2. Now provide the circuit with power. Take the power supply and set the voltage to 9 volts. Connect the positive wire to female edge connector ST1 and connect the ground wire to female edge connector ST7. Then plug the power supply into a socket so that the circuit has power.

3. The circuit is now ready to transfer the required data. Start the appropriate batch file (in our example, that would be *BURNER2.BAT)* from its directory on the companion CD-ROM. After following the instructions that appear on the screen, the batch file will execute and burning is complete.

4. Now disconnect the power supply from the circuit and from the socket. Then remove the plug from the serial port. Carefully remove the controller chip from the IC socket and store it in a safe area until you need it for the next circuit.

# Chapter 29:
# The Model Train Lighting System

# Chapter 29: The Model Train Lighting System

The circuit in this chapter has quite a bit to offer as far as functions go. It also requires some time to build. In addition, you must consider several items and you'll have to perform tasks that we haven't yet explained.

## The Circuit And Its Function

The model train lighting system is a circuit for controlling lights or other electric appliances in a model train set. Besides lights, you can use this circuit to control switches, gates or similar appliances.

Basically, the circuit gives you the option of switching eight times eight (64) outputs on and off independently from one another using the software running on the computer. You can connect small lights, gates and switches to these outputs using wires.

## The Foil Side Of The Board

Unlike previous circuits and chapters, we'll first talk about the foil side of the board. This is because the procedure for building this circuit is different from the preceding chapters.

Until now you have made the connections of the foil side manually on a standard European ("pre-perfed") circuit board with wire. In this case, due to the many components and the high number of necessary connections between the components, this is not possible. A standard European ("pre-perfed") circuit board doesn't have sufficient room. So, we'll have to find a different way.

## The etched board

To solve the space problem, you need to use a different method for creating conducting paths. Since you cannot install the individual wires closely enough together, we need to create the connections in such a way that space is saved. The best possibility of achieving this is to create freely definable conducting paths. We'll circumvent the limitations caused by the arrangement of holes on a standard European ("pre-perfed") circuit board.

To create these freely definable conducting paths, we'll use a procedure that you cannot, as a rule, perform yourself. In this procedure, the required conducting paths including their solder eyes are first created like a drawing on a piece of paper or a foil.

Then you have to transfer these connections to a specific board. This board is similar to the circuit board you are familiar with, however, it doesn't contain any holes. Also, the copper layer does not consist of just rings, but instead, covers an entire side of the board.

If the two required items, the conducting path pattern and the copper-coated board, are present, then we will transfer the conducting paths from the pattern to the board, so that only the specified conducting paths remain as a copper layer. To do this, we'll need a pattern, which we can have etched into the board.

Since this procedure is quite time-consuming and labor-intensive, and can only be performed under certain conditions, we're giving you this information only to serve as a brief explanation. You don't have to etch the board yourself. However, if you have experience in this type of work, you're welcome to do it yourself.

Ordinarily, you will be able to have this work done at your electronic dealer. If your dealer cannot etch a board for you, he or she can surely name a company that can do the work.

## The layout pattern

To etch the board, a company requires an appropriate pattern for the conducting paths, called the board layout. You don't have to draw this yourself. We printed a drawing of the foil side of the required board for you in this book.

We printed the drawing mirror-inverted. It's as if you were looking from above at the component side and then through to the foil side. Use the numbers and letters in the drawing to get your bearings. Ultimately, the board must be etched in such a way that you can read these characters when you look at the foil side of the board. The inscription

```
LS
8*8 OUT
R.M.5.96
```

must be legible on the board.

The drawing is on a 1:1 scale, i.e., the size on the drawing corresponds exactly to the dimensions you'll need on the board. It is especially important that you adhere to this size. If you transfer the drawing either reduced or enlarged, the IC sockets and the ICs won't fit. Small changes in the distances between holes can result in the components no longer fitting.

So you have a pattern for the conducting path arrangement. However, you still need to solve the problem of making the pattern available to the company that is going to make the board for your. You have several options.

1. Your first option would be to cut the drawing out of the book. We don't recommend this solution, since it involves damaging the book.

2. To leave the book intact, we recommend another solution: You can copy the drawing out of the book. However, this solution has another problem. The copy must be exact and of good quality. If not, the distances we were talking about earlier won't be correct, or something worse could happen. If the copy is dirty, the additional marks or dirty areas will cause unintentional connections.

   Remember that everything that is on this sheet will also be transferred to the board, including things that are so small you can hardly see them. This can result in contacts on the board between conducting paths that influence the whole function of the board. On the other hand, copying can also result in interruptions between conducting paths that will also be transferred to the board. Even if the interruptions are minimal and you can hardly see them, they can still lead to trouble.

3. Your best option is to send the drawing to the company in digital form. That means sending the drawing as a file. Naturally, the company must be able to work with this file and use it as a layout pattern. We've saved the drawing as a file in PostScript format on the companion CD. You will find it in the *Chap_29* directory under the name *Modell.ps*. Copy the file to a diskette, send it to the company and have them etch a board from the layout file.

**Important Tip**

Once you have the board etched, the necessary conducting paths are present. What you don't have are the holes for soldering the components. If the company that etches the board will also drill the holes, by all means, take advantage of this service. If you bore the holes yourself, be careful. If possible, use a stationary drill press. Boring the holes by hand often does not yield the desired results. You can quickly damage a conducting path, or you might drill the holes at an angle, so that it's not possible to insert the components smoothly.

As you have found out, creating an etched board is not so easy. However, once you have the etched board, the remainder of the work is relatively easy, since you don't have to do the time-consuming wiring.

## Building The Circuit

Once the board is finished and has the appropriate holes, the next step is to put the components on it. This means buying the necessary components and then soldering them onto the specified positions on the board.

**Important Tip**

When placing the components, make sure you don't place them in the wrong holes. This can easily happen if a hole didn't get bored. In the long run, you will notice a missing hole, but severing an existing connection represents work and is a potential source of errors. Work very carefully, and don't rush when attaching the components.

Unlike the foil side, the component side of this circuit hardly differs from the ones on the circuits you've dealt with up to now. Only the orientation on the board isn't quite so easy, since the holes don't form a pattern. You can't use the method of counting the holes off and then inserting the components into the correct holes. However, this won't be a problem with this board. This next drawing shows the layout of the individual components.

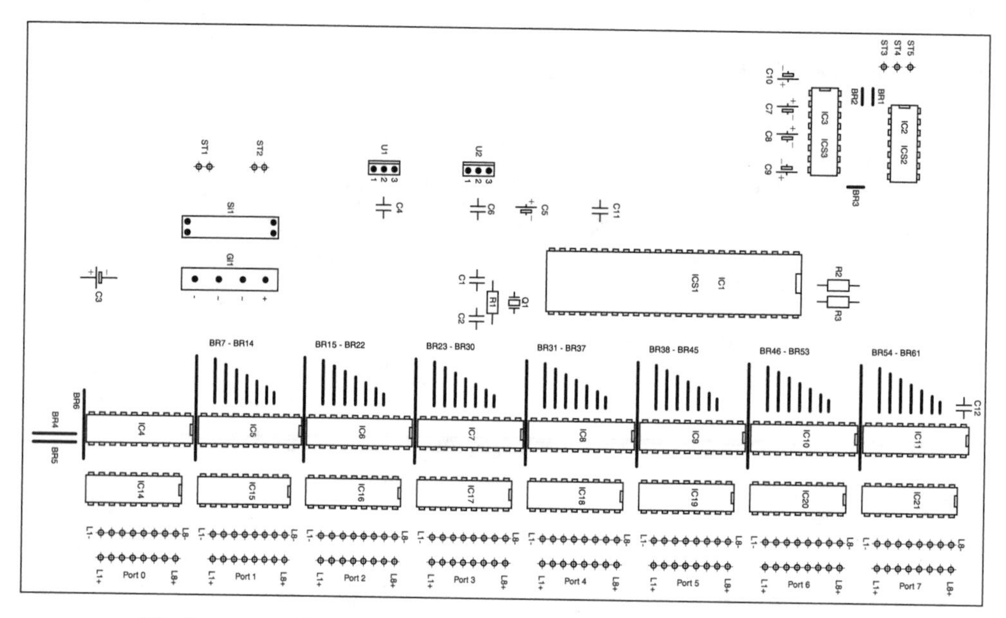


*The layout of the components on the component side of the board*

## Parts list

Refer to the parts list in the following table to determine which components are used in this chapter.

| Parts List | | |
|---|---|---|
| Label | Name | Component type |
| BR1 - BR62 | Conducting path bridges | Plastic-coated copper wire |
| R1 | 1 MΩ | Resistor |
| R2, R3 | 10 kΩ | Resistor |
| C1, C2 | 22 pF | Condenser |
| C3 | 4700 µF 25 V | Condenser |
| C4, C6, C11, C12 | 100 nF | Condenser |
| C5 | 220 µF 16 V | Condenser |
| C7 - C9 | 1 µF | Condenser (Tantalum) |
| C10 | 10 µF | Condenser (Tantalum) |
| U1 | 78S08 + small heat sink | Voltage regulator |
| U2 | 78S05 + small heat sink | Voltage regulator |
| GL1 | Rectifier 8 A/50 V | Rectifier |
| S1 | Fuse holder including fuse 8 A | Fuse |
| Q1 | 8.000 MHz | Quartz Mini Crystal |
| IC1 | MC 68HC11A1 | IC, programmed by the burner circuit of previous chapter |
| IC2 | SN 74HC08 | IC |
| IC3 | MAX 232 | IC |
| IC4 bis 11 | 74HC573 | IC |
| IC14 bis 21 | ULN 2803A | IC |
| ICS1 | IC socket 48-pin | IC socket |
| ICS2 | IC socket 14-pin | IC socket |
| ICS3 | IC socket 16-pin | IC socket |

After getting the necessary components, put them on the board and solder them. There's nothing special to mention about the components, merely the fact that the voltage regulators U1 and U2 must each be provided with a small heat sink. You screw the heat sinks on the rear side of both components. They prevent the components from overheating.

With the controller chip, you have to load the necessary control information into the chip using the burner circuit of the previous chapter. Only then can the controller do its work properly in this circuit.

**Important Tip**

Normally all the required holes on the board will be the same size. It's possible that the rectifier or the holder for the fuse won't fit into the holes because they aren't big enough. In this case, you'll have to use a larger drill to expand the holes. Make sure you don't drill the holes too large, or you could destroy the conducting path or the solder eye on the foil side of the board.

Build the board following the specifications. The finished product should look like the following figure.

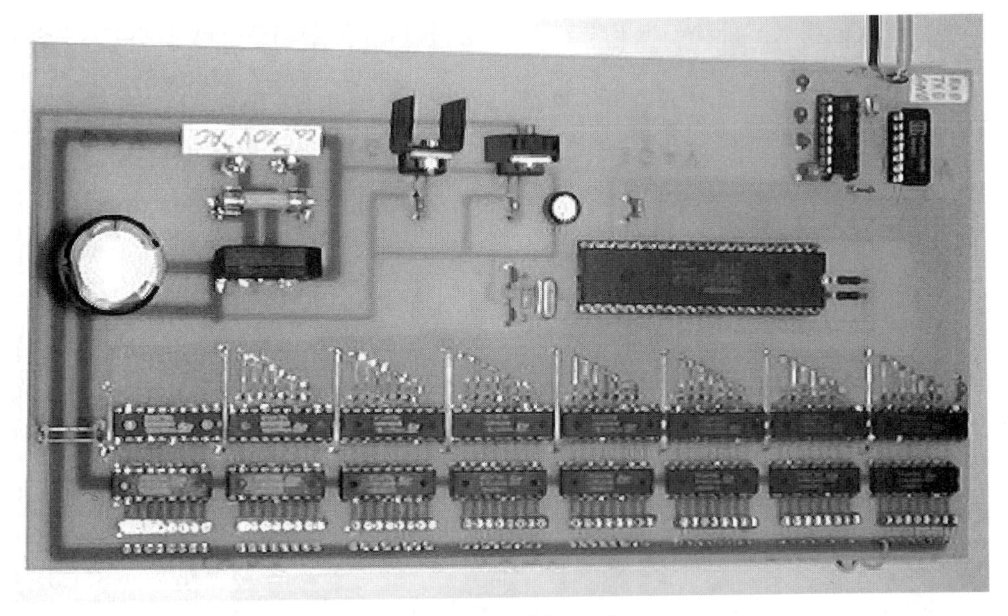

*The finished board*

## Additional Work On The Circuit

After you finish soldering on the components, there are still two important items that are absolutely necessary for the function of the circuit.

### The power supply

The circuit requires its own power supply. Up to now you have dealt with circuits that had an additional external power supply. With this circuit, we took a different route.

The circuit requires alternating-current voltage to execute its function. Up to now, this book hasn't required any electronic knowledge. We're going to keep it that way. We'll just mention that the circuit is designed so that the necessary direct-current voltage is generated in the circuit itself, supplying both the components on the board as well as the lights connected to the outputs with power.

For this purpose, it is necessary to reduce the 120 volts of alternating-current voltage that comes from the socket to approximately 9 volts of alternating-current voltage using a transformer. These 9 volts are enough for the circuit to perform its necessary functions. But, how do you get the 9 volts?

It's actually not very difficult; however, there's something we need to discuss.

**Important Tip**

The tasks we are going to discuss for getting 9 volt alternating-current voltage should be performed by a knowledgeable electronics dealer, since it is necessary to work with 120 volts of alternating-current voltage. People who don't know what they are doing run the risk of injuring themselves. Since only a few tasks are necessary, take the resulting expenses as part of the bargain and turn to an expert who will perform the tasks for you properly.

What you need is comparable to the power supply you've used with some of the preceding circuits. Only the insides and the output voltage of this "power supply" are different.

The foundation of this power supply is the transformer we mentioned earlier. You require a transformer that is equipped for a primary voltage of 120 volts. The secondary voltage, that is, the output voltage must amount to 9 volts. These specifications are well-known in the electronics shop, so they can help you make the right purchase.

It's important to get the right size, i.e., the required performance. The more lights you wish to operate simultaneously and the higher the performance per light, the greater the performance of the transformer.

## Building the power supply

The transformer has two connection pairs. You solder the power cable to the transformer input pair, with a fuse holder and fuse in one of the two power lines. The fuse must be selected to match the performance of the transformer. The power cable has a plug for a normal 120 volt socket on one end. You can buy these cables at an electronics shop.

You can easily build the transformer in conjunction with the fuse holder and the feed line on a standard European ("pre-perfed") circuit board. Then connect the other connection pair (the transformer output) to a two-conductor cable, which you furnish with male edge connectors and plug into pins ST1 and ST2 of the circuit. It doesn't matter which cable goes to which pin in this case.

**Important Tip**

120 volts are connected with the transformer. Have an expert create the power supply, and have him furnish it with a protective casing, so that voltage conducting parts are no longer accessible. Don't plug the power cable into the socket until the entire power supply has been securely built. Do not plug the power cable into the socket if any of the voltage conducting parts can be touched. It's impossible to completely rule out shock accidents where you come into contact with such areas.

There are cases which have a connection plug for a socket integrated, similar to what you are familiar with from normal power supplies. If you build the transformer into this kind of case, the power wire is no longer necessary. The required connections are installed only within the cases, so that they cannot be touched from the outside. This increases safety.

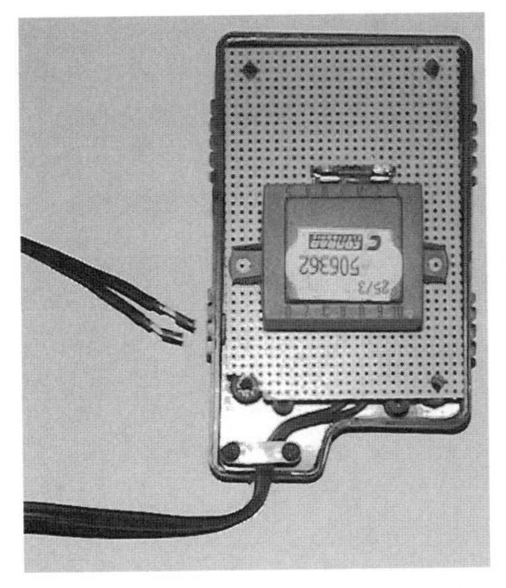

*The power supply*

## The cable to the computer

Once you have a secure power supply, the circuit still requires connection to the computer. Since you control this circuit through the serial port, a two-conductor shielded cable is sufficient. It's important that the cable be shielded, since the ground connection takes place here. Don't cut the cable too short so you can place the circuit where you want it to be.

1. Connect the one end of the cable to the circuit using three male edge connectors. On the other end, solder a 25-pin Sub D connector (female), which then plugs into the socket of the serial port.

2. Take the cable and determine which conductor is wire 1. You should be able to clearly identify this on both ends of the cable. As a rule, the conductors are in different colors, so it shouldn't be difficult. The other conductor is wire 2. The schield is referred to as the ground.

3. First solder wire 1 to contact 2 of the connector. Since the connector is numbered, you should be able to find the right contact.

4. Next, connect wire 2 to contact 3 of the connector. Twist the schield into a wire and connect it with contact 7.

5. You'll also need two bridges. Each of these connect two contacts of the connector. Place a bridge from contact 4 to contact 5. Place the second bridge from contact 6 to contact 20. When soldering, make sure that the solder doesn't drip and create any unintended connections to other contacts or between the conductors.

6. You can also place a case around the connector if you wish. However, think about tension relief for the cable.

7. On the other end of the cable, solder a male edge connector to each conductor. Twist the shield together and solder a male edge connector here as well. Finally, plug wire 1 into pin ST5 and plug wire 2 into pin ST4 of the circuit. Plug the ground wire into pin ST3.

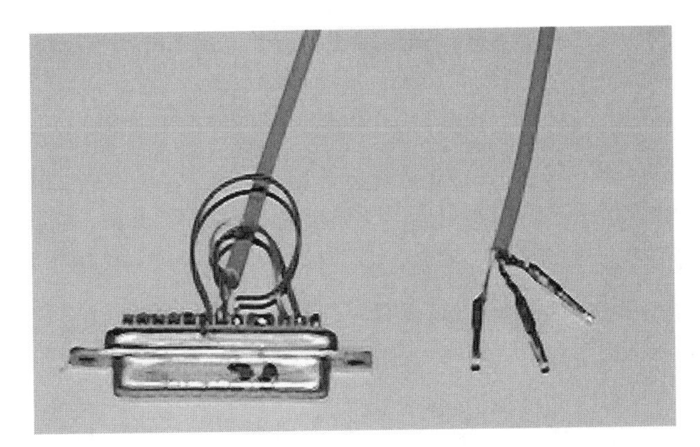

*The connection cable to the computer*

You have now performed all the necessary tasks for the circuit.

## Using The Circuit With The Companion CD-ROM

You'll find the program for controlling the circuit on the companion CD-ROM in the *Chap_29* directory. If you wish to modify the software, use the source code files in the *Source* subdirectory.

When you run the program, the following screen appears.

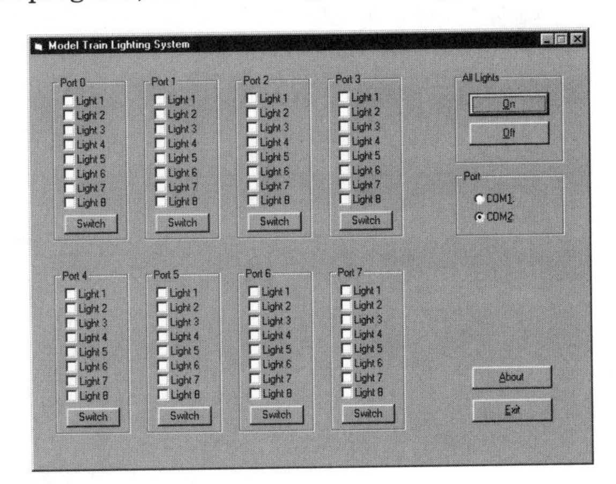


*The program for controlling the model train lighting system*

At the right border of the window you will find the option for selecting the port to which you have connected the circuit. By default, COM2 is selected. If you are running the circuit at COM1, make the necessary correction.

You will find two additional buttons above the port area. Click the *On* button to enable all the outputs of the circuit, i.e., all the lights are switched on. The *Off* button gives the opposite effect. All the lights, i.e., all the outputs are switched off.

The two buttons in the lower right corner of the window let you display brief information about the program or exit the program.

The *Port* areas, which cover almost the entire window, are important. You can use check boxes to make an exact choice as to which lights from which Port will be enabled/disabled. A checked check box means an active outlet. However, the changes you make to these settings don't go into effect until you have clicked the *Switch* button of the appropriate *Port* area.

## Using The Circuit

The board contains eight ports, which each have eight outputs. Each output is made up of both a positive and a negative or ground contact. There are 64 contact pairs to which you can connect the lights of the model train.

To use the circuit, follow this procedure.

1. First, lay the wires for the lights so that they all end at the board. You can connect 12 volt lights to the outputs, which can be supplied with a maximum of 500 mA of current.

**Important Tip**

Since the board is designed so that all contact pairs have a common positive terminal, you have the option of laying only eight signal wires in a specific area of the model train set with a common positive wire, which is then divided in this area and placed alongside the individual lights.

2. As soon as all the desired wires to the lights are installed and connected with the contacts or the contact pairs, make the connection to the computer. Then start it up. Only then do you provide the circuit with power, plugging the transformer feed line to the into the socket.

3. The circuit is now ready to use. Start up the software program and make the desired settings.

## Watch out for the following

The circuit is designed for a total of 64 connections. You can imagine that a great deal of current and power will be used if all 64 outputs are loaded simultaneously with the maximum utilization. In such a case, it is possible that the individual elements of the circuit could heat up tremendously, including the conducting paths of the board. You can tell by the width of the different paths that this circumstance has been taken into account. However, in such occurrences you need to take precautionary measures.

Here's what to do: solder additional wires on the wide conducting paths that you are familiar with from the standard European ("pre-perfed") circuit boards. Also, use lights that require less, rather than more, power.

# Index

# I



# L



# M



# P



# R



# S

# T



# V

Step-by-Step books are for users who want
to become more productive with their PC.

## Building Your Own PC

Windows 95 and today's new, exciting applications demand a fast computer with lots of horsepower. The new Pentium-class computer systems deliver that power with room to grow. The price for a Pentium system has continued to drop to a level that makes a new computer system very affordable. But you can save even more money by building your own computer.

Building a computer is not the hassle that it was just a few short years ago. Today you can buy "standard" off-the-shelf components everywhere. Building a personal computer might be better called assembling a computer since most of the parts are very reliable and interchangeable.

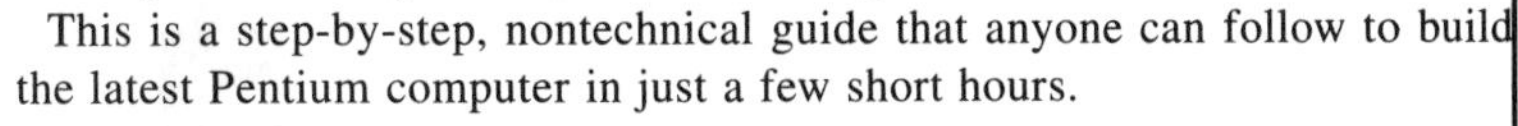

This is a step-by-step, nontechnical guide that anyone can follow to build the latest Pentium computer in just a few short hours.

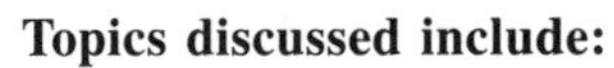

**Topics discussed include:**

- Selecting a CPU
- Deciding on a motherboard
- Which bus—PCI or VLB?
- Adding speedy cache memory
- Configuring the motherboard and more!

Author: Arnie Lee
Order Item #B320
ISBN: 1-55755-320-3

Suggested Retail Price
$24.95 US/$32.95 CAN

**To order direct call Toll Free 1-800-451-4319**

In US & Canada Add $5.00 Shipping and Handling
Foreign Orders Add $13.00 per item. Michigan residents add 6% sales tax.

Visit our website at www.abacuspub.com

## CD-ROM Installation

This CD-ROM contains all the programs you need with the completed circuit boards. The directories are titled by chapter. For example, Chapter 8 equals chap_08 on the CD, Chapter 9 equals chap_09 and so on.

There are two ways of copying and installing the programs. You may run the SETUP.EXE program located on the root of the CD-ROM. This will install all the project's programs including the source code into a directory on your hard drive. You will find that in each directory (labeled chap_#) several files. They include the .EXE (executable) and source files (.FRM, .BAS and .VBP). The dynamic link library is also included in each folder (portio.dll). Or, you can simply copy the projects one at a time using Windows Explorer.

You may also install the fantastic shareware program PC TRACE located in the PCTRACE directory. Simply click **Start**, select **Run...** and type D:\PCTRACE\SETUP.EXE. The program will be installed on your PC. PC Trace was written by Ehlers Technical Consultants, 4520 S. 58th St., Lincoln, NE 68516. This product is shareware and if you use the programs, you are required to purchase a license to continue using them. Be sure to read the README.WRI in order to comply with the agreement.

## Running SETUP.EXE

Click **Start**, select **Run...** and type D:\SETUP.EXE. You will see the following screen.

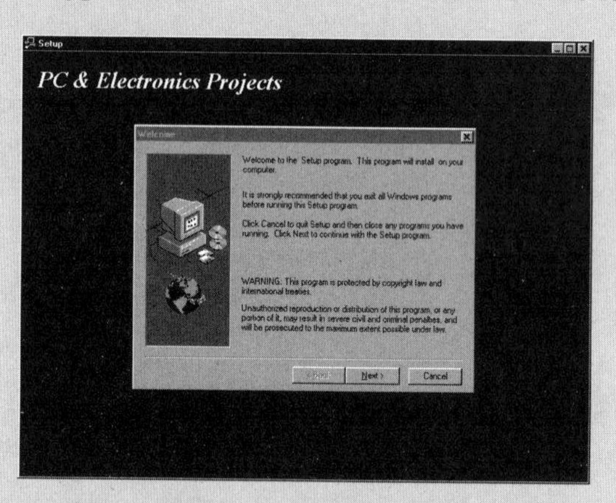


This is the Welcome Screen; click the [Next >] button to continue.

Follow the prompts and after installation you can load and run each of the project programs. Simply click **Start**, select **Run...** and locate the project you want to run.

**See Chapter 6 for more information on the Companion CD-ROM**